MANGANESE REMOVAL FROM GROUNDWATER: ROLE OF BIOLOGICAL AND PHYSICO-CHEMICAL AUTOCATALYTIC PROCESSES

MANGANESE REMOVAL FROM GROUNDWATER: ROLE OF BIOLOGICAL AND PHYSICO-CHEMICAL AUTOCATALYTIC PROCESSES

DISSERTATION

Submitted in fulfillment of the requirements of
the Board for Doctorates of Delft University of Technology
and
of the Academic Board of the UNESCO-IHE
Institute for Water Education
for
the Degree of DOCTOR
to be defended in public on
Tuesday, 28th June 2016, at 15:00 hours
in Delft, the Netherlands

by

Jantinus Henderikus BRUINS
Master of Environmental Science, Open University Heerlen
born in Eext, the Netherlands

This dissertation has been approved by the
promotor   : Prof. dr. ir. M.D. Kennedy and
copromotor   : Dr. ir. B. Petrusevski

Composition of Doctoral Committee:

Chairman   Rector Magnificus TU Delft
Vice-Chairman   Rector UNESCO-IHE
Prof. dr. ir. M.D. Kennedy   UNESCO-IHE / TU Delft, promotor
Dr. B. Petrusevski   UNESCO-IHE, copromotor

Independent members:
Prof. dr. ir. W.G.J. van der Meer   Delft University of Technology
Prof. dr. ir. J.P. van der Hoek   Delft University of Technology
Prof. dr. V. L. Snoeyink   University of Illinois, USA
Dr. P.S. Hofs   Evides Water, the Netherlands

Prof. dr. ir. M.E. McClain   TU Delft / UNESCO-IHE, reserve member

Published by:
CRC Press/Balkema
PO Box 11320, 2301 EH Leiden, the Netherlands
Pub.NL@taylorandfrancis.com
www.crcpress.com – www.taylorandfrancis.com
ISBN 978-1-138-03002-2

Acknowledgements/dankbetuiging

Geheel tegen de schijnbare gewoonte in bij het verwoorden van een dankbetuiging, wil ik niet eindigen, maar beginnen met diegenen, die vóór en tijdens mijn studie een zeer belangrijke rol hebben gespeeld. Het noemen van de partner zal voor velen een cliché zijn, maar ik kan u verzekeren hoe anders dit is als het je zelf betreft. Want het was zeker niet alleen tijdens deze studie dat Martha mijn steun en toeverlaat is geweest en nog altijd is. Al sinds onze eerste kennismaking in 1984 heeft ze me niet anders meegemaakt dan studerende. Ook toen Marcel en Ronald onderdeel uit gingen maken van ons gezin, was zij het die verreweg het grootste deel van de opvoeding en zorg voor onze zonen op zich heeft genomen. Als zeer toegewijde moeder, heeft zij zich hierdoor volledig weggecijferd en daarmee haar werkzame carrière opgeofferd. Dit alles om mij de kans te geven me te ontplooien en te komen waar ik nu sta. Martha, dit verdient alleen maar mijn diepste respect, waar ik jou - uit de grond van mijn hart - enorm dankbaar voor ben.

Ik noemde al Marcel en Ronald. Mijn "leven lang leren" heeft er ongetwijfeld voor gezorgd dat ik er niet altijd voor jullie was, terwijl dit wel zo had moeten zijn, zoals jullie dat ook van een goede vader hadden mogen verwachten. Dit spijt me, maar ik troost me met de gedachte dat dit gemis, ruimschoots is gecompenseerd door jullie fantastische moeder.

Nog meer aan de basis van mijn carrière als "eeuwige" student, stonden mijn ouders. Zij hebben mij altijd gestimuleerd, waar en waarmee ze ook maar konden om een opleiding te volgen. Dit begon al op de middelbare school en werd met evenveel enthousiasme doorgezet tijdens mijn promotietraject. Pap, helaas heb je de voltooiing van dit traject niet mee mogen maken, maar ik ben heel blij dat de start je nog wel gegeven is. Mam, je weet niet hoe ongelofelijk blij ik ben dat jij dit slotstuk wel mee kunt maken. Dank voor alles wat jullie voor mij gedaan hebben en ik hoop dat je een beetje trots kunt zijn. Het is dezelfde trots, die ik ook bespeurde bij mijn schoonouders. Helaas heeft ook mijn schoonvader het einde van mijn promotietraject net niet mogen meemaken.

Natuurlijk was het uitvoeren en afronden van mijn promotietraject een stuk moeilijker, zo niet onmogelijk geweest als mijn werkgever WLN (verpersoonlijkt door Hilde en Gerrit) en hiermee ook de twee moederbedrijven; Waterleidingmaatschappij Drenthe en Waterbedrijf Groningen mij niet de ruimte en de financiële steun hadden gegeven die ik kreeg. Ik hoop dat dankzij dit onderzoek, voor hen en met hen de hele drinkwaterwatersector, de klassieke grondwaterzuivering en in het bijzonder de opstart van het ontmanganingsproces in de toekomst op een snellere en nog duurzamere wijze kan worden uitgevoerd.

Mijn promotieonderzoek werd uiteindelijk uitgevoerd bij UNESCO-IHE in Delft. Branislav, ik ben je dankbaar voor je inzet, stimulans en immer positief kritische houding als het ging over het werk dat ik uitvoerde, de "abstracts" en "papers" die ik moest indienen. En natuurlijk vergeet ik zeker niet de persoonlijke rondleiding door Belgrado, die ik samen met Salifu en Valentine van jou en je echtgenote heb gehad rond een congres. Dit blijft een mooie en bijzonder waardevolle herinnering. En in de loop van de jaren kregen onze gesprekken ook vaak een persoonlijke(re) noot. Yness, thank you very much for everything you have done for me. Your experience in writing a good paper for a peer reviewed journal was very valuable for me, and I know that you had to stay calm and patience over and over again, to learn me a few of your skills. Your creativity was also very helpfull. Tijdens het schrijven van artikelen, maar ook voor dit proefschrift heb ik de nodige adviezen van Joop gehad. Jouw jarenlange ervaring heb ik direct kunnen omzetten in een enorme kwaliteitsimpuls wat betreft de manuscripten. Tot slot heeft Maria als promotor een belangrijke stempel gedrukt op dit onderzoek door richting te geven, maar vooral ook door grenzen te stellen. Als ze dit laatste niet had gedaan, was ik waarschijnlijk nog steeds bezig geweest met praktisch onderzoek.

Het gehele promotietraject is dan wel uitgevoerd "onder de vlag" van UNESCO-IHE, maar het grootste en ook het belangrijkste deel van het praktische onderzoek is uitgevoerd bij de collega's van het Vlaamse waterbedrijf Pidpa. De directie en medewerkers van Pidpa ben ik dan ook zeer erkentelijk voor hun steun en toewijding tijdens mijn onderzoek. Hoewel binnen Pidpa velen een bijdrage hebben geleverd wil ik toch een paar collega's in het bijzonder noemen. Hun bijdrage is cruciaal geweest in dit geheel. Het balletje is natuurlijk gaan rollen door de goede contacten die ik, sinds het laatste decennium van de vorige eeuw, heb met Koen. Op het moment dat mijn promotieonderzoek en het onderwerp ervan ter sprake kwam, was Koen meteen enthousiast en had wel een locatie beschikbaar voor onderzoek. Dit aanbod heb ik uiteraard in dank aanvaard en met hulp van mijn voormalige collega Chris, mijn huidige collega Karl (toen Vitens), "de andere" Koen van Pidpa en een aantal medewerkers van de locatie Grobbendonk, werd de pilot opgebouwd. Bij de begeleiding/advisering van dit pilotonderzoek heb ik enorm veel steun gehad van beide Koenen. Voor een groot deel werd dit pilotonderzoek ook door hen uitgevoerd, uiteraard met hulp van Martine en Ann.

Het waren overigens niet alleen mijn collega's Chris en Karl van WLN die een bijdrage aan mijn promotietraject hebben geleverd, maar met hen vele anderen. Natuurlijk mijn directe collega's van technologie, die door mijn afwezigheid, misschien af en toe "een stap harder moesten lopen", maar die daarentegen ook vele dagen niet mijn "gezeur en geklaag" hoefden aan te horen. Gerhard (ook als mede auteur), Marsha, Pim en Arnout hebben veel werk verzet tijdens het deel waar de moleculaire microbiologie om de hoek kwam kijken. Verschillende laboratorium collega's van de diverse afdelingen heb ik mogelijk schrik aangejaagd toen ik zelf laboratoriumanalyses kwam uitvoeren. En natuurlijk bedank ik ook mijn collega's, die tijdens het uitvoeren van hun werkzaamheden vaak niet in de gaten hadden dat ze een bijdrage leverden aan mijn onderzoek. Veel van de te onderzoeken monsters zijn aangeleverd door Arjan (en zoals je ziet Arjan, mijn dankwoord richting jou heeft een vergelijkbaar lettertype gekregen als het dankwoord aan een ieder........dat valt niet tegen toch?).

Voor een aantal specialistische onderzoeken ben ik geholpen door verschillende deskundigen. Het onderzoek met de Electron Paramagnetic Resonance (EPR) heb ik zelf mogen uitvoeren bij de Rijks Universiteit Groningen, na een "spoedcursus EPR" door Dr. Wesley Browne. Onderzoeken met XRD werden uitgevoerd door Ruud Hendrickx van de TU Delft. Ton heeft bij Wetsus heel veel onderzoek gedaan met Raman spectroscopie en Jelmer heeft tegen het eind van mijn promotieonderzoek een aantal SEM-opnamen gemaakt. En hoewel deze opnamen prima waren, had ik toch heel graag gehad dat ook deze laatste SEM opnamen, net als alle andere, gemaakt waren door Arie. Het heeft helaas niet zo mogen zijn. Arie, ik koester de fijne uren samen achter de SEM. De gesprekken over het werk, de vakanties, de foto's en onze families. Maar we begonnen natuurlijk met koffie, want zoals je altijd zei: "*dat apparaat moet toch eerst even opwarmen*". Jouw werk zal herkenbaar blijven in het onderzoek van velen en op deze wijze hoop ik de herinnering aan jou ook voor altijd te laten voortleven. Arie, bedankt voor wie je was en wat je voor mij hebt gedaan!

Vanuit het "werkveld" werd mijn promotietraject van de nodige ondersteuning voorzien, door een begeleidingsgroep van experts. Hiervan maakten Koen en Joop ook deel uit en verder bestond de groep uit: Ans (RIVM), Bas (KWR), Ben (De Watergroep) en Jacques (Vitens). Een ieder heeft hier op zijn of haar wijze een waardevolle en belangrijke bijdrage geleverd aan het geheel. Met Jacques reisde ik vaak samen naar Delft en tijdens de 'lange' reis hadden we vaak boeiende en interessante gesprekken, waardoor de afstand Zwolle - Delft soms juist wel heel erg kort leek. Van KWR hebben ook Dirk en Bart (beiden als mede auteurs) een waardevolle bijdrage geleverd. Ook verschillende collega's van Nederlandse, Vlaamse en Duitse waterbedrijven hebben informatie verstrekt die belangrijk was voor de inventarisatie van "ontmanganing in de praktijk". Dit heeft uiteindelijk heeft geleid tot de 1e publicatie.

Behalve dat ik zelf de nodige onderzoeken heb uitgevoerd, zijn er ook een aantal MSc studenten geweest die een deel voor hun rekening hebben genomen. The first student was Edilberto Ayala Baquero from Colombia and student at Wetsus. Chris was de 2e student en hij volgde een opleiding aan Van Hall-Larenstein. Next students were all from UNESCO-IHE: Ahmed Abdullah Saif Al-Abri and Younis Sulaiman Hamed AlZakwani, both from Oman, Nicholaus Angumbwike Njumbo from Tanzania and Clement Ndungutse from Rwanda. De student die het geheel afsloot was Nicolas Soenens van de Universiteit van Antwerpen.

Tot slot wil ik een ieder bedanken, die ik niet heb genoemd in dit dankwoord, maar die toch een belangrijke bijdrage heeft geleverd aan het welslagen van mijn promotieonderzoek.

BEDANKT   **THANK YOU**

Jantinus Bruins
Juni, 2016
Assen, Nederland

Voor de twee belangrijkste vrouwen in mijn leven.

For the two most important women in my life

Summary

Groundwater is the predominant source of drinking water globally. However, untreated groundwater contains compounds that are undesirable in drinking water, such as methane, ammonia, iron and manganese.

The presence of manganese in drinking water is detrimental, because of health, aesthetic and practical reasons. Increased levels of manganese in drinking water are associated with neurologic symptoms, especially in young children. Moreover, a synergistic toxic effect exists in combination with arsenic.

In European countries, health related effects caused by manganese in drinking water are negligible and the main problems are of aesthetic and practical concern, particularly in groundwater treatment.

In The Netherlands, manganese is removed by conventional groundwater treatment consisting of aeration and rapid (sand) filtration. Such a treatment process is easy to operate, cost effective and sustainable as it does not require the use of strong oxidants such as O_3, Cl_2, ClO_2 and $KMnO_4$ with the associated risk of by-product formation and over or under dosing. However, the application of aeration-filtration has also some drawbacks. In particular, the long ripening time required for filter media is a major concern. Currently, water supply companies have to waste large volumes of treated water, which reduces the sustainability of the process. In addition, the costs associated with filter media ripening (man power, electricity, operational and analyses costs) are high. Consequently, this thesis describes the removal of manganese from groundwater by microbial and physico-chemical auto-catalytic processes, with a specific focus on revealing the mechanisms involved in filter media ripening with the aim to shorten or even completely eliminate filter media ripening.

Initially, data was collected from over 100 full-scale groundwater treatment plants, mainly in The Netherlands, Belgium and Germany and an inventory of the most important (water quality and operational) parameters required for successful manganese removal by aeration-rapid sand filtration, was made. Univariate statistics and assessment of available data (from over 100 plants) indicated that a very effective manganese removal efficiency in the first aeration-filtration stage with simultaneous removal of iron and ammonia, could be achieved under the following conditions:

- NH_4^+ removal efficiency   :   > 85%
- iron loading per filter run   :   < 2.7 kg Fe/m^2
- pH of filtrate   :   > 7.1
- filtration rate   :   < 10.5 m/h
- empty bed contact time   :   > 11.5 min
- oxygen in filtrate   :   ≥ 1 mg/L

An important step in revealing the processes and mechanisms involved in ripening of the filter media was the identification of the manganese oxide present. To characterize and identify the manganese oxide present on filter media coatings, the following techniques were used: Röntgen diffraction (XRD), Raman spectroscopy, Scanning Electron Microscopy and Energy Dispersive X-ray analysis (SEM-EDX) and Electron paramagnetic resonance (EPR). With these techniques, an amorphous type of manganese oxide: Birnessite was identified in all samples examined. Birnessite has excellent properties to adsorb ions, such as Mn^{2+}, and has highly auto-catalytic oxidative properties. Both of these properties make Birnessite extremely suitable to remove Mn^{2+} from groundwater in an effective and efficient way.

The next part of the research focused on investigating the process of filter media ripening. Therefore, pilot tests were conducted at the water treatment plant Grobbendonk (Pidpa, Belgium). During the ripening process, samples of the filter media as well as the backwash water were collected and analyzed. Applying the aforementioned techniques, the formed manganese oxide was characterized and identified. Moreover, with the combination of SEM and EPR it was possible to distinguish whether the Birnessite was formed biologically or physico-chemically. These techniques showed that at the beginning of the ripening process the produced Birnessite was of *biological* origin. As filter ripening progressed and a coating developed on the filter media, the produced Birnessite became predominantly of *physico-chemical* origin. After approximately 500 days, all Birnessite present on the filter media was produced physico-chemically.

Throughout the whole ripening period manganese oxide particles collected from filter backwash water were consistently of biological origin, suggesting that biological oxidation of adsorbed manganese took place throughout the filter run. Consequently, the bacteria population present in (freshly ripened) manganese removal filters was examined. For this purpose, the following molecular DNA analyses were used: "next generation DNA sequencing", qPCR and MALDI-TOF MS analysis.

The "next generation DNA sequencing" analyses, showed a bacteria population shift during the start up phase of the manganese removal process. In the filtrate of the iron removal filter as well as in the feed of the manganese removal filter, the iron oxidizing genus *Gallionella* was dominant (> 97% of the total bacteria population), whereas the backwash water of the manganese removal filter comprised only 12.4% of the genus *Gallionella*. Bacteria of the genus *Nitrospira* and the genus *Pseudomonas*, 25.7 % and 14.3%, respectively, were also present in the manganese removal filter. However, 47.6% of the bacteria population in the manganese oxidizing column, consisted of small groups of bacteria which remained unknown.

Nitrospira is known to be involved in the oxidation of nitrite to nitrate and therefore its presence is expected, because in addition to manganese, ammonia was also oxidized in this filter. *Pseudomonas* sp. and in particularly *P. putida* is known to be capable of oxidizing Mn^{2+}.

However, qPCR established that the presence of *Pseudomonas putida* was very limited. Less than 0.01% of the genus *Pseudomonas* present was of the species *Pseudomonas putida*. After successive culturing, some strongly related *Pseudomonas* species (amongst others: *P. gessardii, P. grimontii and P. koreensis*) were identified with MALDI-TOF analysis. At GWTP Grobbendonk, *Pseudomonas* sp. is most likely the manganese oxidizing bacterium genus playing an important role in the start up phase of filter media ripening. However, it is not known whether this bacterium genus is operating alone or as part of a microbial consortium.

A study was carried out to assess the potential of manganese oxide-coated filter media (MOCS/MOCA) to reduce the ripening time of filters with virgin media. The addition of a layer of fresh MOCA to the filter eliminated the ripening time completely, while a layer of dry MOCS introduced to a virgin sand filter did not significantly affect the ripening period. Both tests were applied at different locations, with different operational and water quality parameters.

Based on the aforementioned findings, a pilot study with fresh and dry coated filter media was performed under controlled conditions. Virgin filter media (sand and anthracite) were used as reference materials. From this study it was concluded that, with comparable process conditions, the duration of filter media ripening with virgin sand and anthracite was similar.

Furthermore, it was shown that freshly prepared manganese oxide coated filter media has excellent properties to enhance the ripening process and was capable of *eliminating* ripening time completely. On the other hand, the use of dry MOCS, which adsorbs Mn^{2+} temporarily, did not have a significant impact on the ripening time.

The effect of backwash frequency on filter media ripening was examined at pilot scale. This study confirmed that more frequent filter backwashing negatively affected filter media ripening time with virgin media. Thus, backwash frequency was shown to be a *key factor* in the start up of new filters for manganese removal. The influence of backwashing became less pronounced as filter ripening progressed, due to the development of a thicker layer of biomass and/or auto-catalytically active Birnessite on the media surface. Backwashing showed very little impact on filter ripening time (manganese removal efficiency) when a layer of fresh MOCA/MOCS was used. The backwash frequency depends on the amount of oxidized Fe^{2+} ($Fe(OH)_3$) which is retained by the filter. Consequently, the iron concentration in the feed water and the iron loading are also key factors influencing the ripening time of manganese removal filters.

Prior to this research, it was believed that manganese removal by conventional aeration-filtration was a physico-chemical process dominated by the formation of the manganese oxide Hausmannite and that microbial processes were also involved. This thesis presents *new information* revealing the mechanisms and processes involved in the start up of filters with virgin media in the removal of manganese from groundwater. It was clearly shown that the predominant manganese oxide contributing to filter media ripening was *Birnessite*. It was also found that Birnessite formed at the start of the ripening process was of *biological origin* while as ripening progressed, Birnessite formed became predominantly of *physico-chemical origin*.

Based on the knowledge presented in this thesis, water supply companies can take measures to optimize the filter ripening process, thereby reducing the ripening time. This can be achieved by creating conditions favouring the growth of manganese oxidizing bacteria, for example by limiting the frequency of backwashing (*e.g.,* by limiting the iron loading of the filter).

Finally, filter media ripening can be *completely eliminated* by the addition of freshly prepared MOCS/MOCA, containing Birnessite, to the filter.

Samenvatting

Grondwater is wereldwijd de belangrijkste bron voor de productie van drinkwater. In grondwater zijn van nature verschillende bestanddelen aanwezig, die in drinkwater niet gewenst zijn, zoals; methaan, ammonium, ijzer en mangaan.

De aanwezigheid van mangaan is ongewenst om gezondheidskundige, esthetische en praktische redenen. Verhoogde concentraties mangaan kunnen vooral bij jonge kinderen aanleiding zijn voor neurologische problemen. In combinatie met arseen kan er sprake zijn van een versterkt toxisch effect. In West-Europa zijn de gezondheidskundige effecten ten gevolge van de aanwezigheid van mangaan in drinkwater verwaarloosbaar en zijn de problemen vooral van esthetische en praktische aard. Deze praktische problemen doen zich met name voor bij de zuivering van het grondwater. Zuivering van grondwater vindt in Nederland vooral plaats door beluchting, gevolgd door filtratie. Dit is een eenvoudige, goedkope en duurzame vorm van waterzuivering, omdat hierbij geen gebruik wordt gemaakt van chemicaliën zoals O_3, Cl_2, ClO_2 en $KMnO_4$ ten behoeve van de oxidatie. Dit type zuivering kent dan ook niet de nadelen die verbonden zijn aan het gebruik van deze sterke oxidatoren, zoals de vorming van bijproducten en over- of onder dosering. Met betrekking tot de verwijdering van mangaan kent de toepassing van deze grondwaterzuiveringstechniek ook een aantal nadelen, waarvan de lange rijpingstijd van nieuw filtermateriaal de belangrijkste is. De rijping van nieuw filtermateriaal, voordat een volledige mangaanverwijdering is gerealiseerd, duurt over het algemeen enkele maanden tot soms meer dan een jaar. Voor waterbedrijven leidt het rijpingsproces tot een verlies van gezuiverd water, additionele kosten (arbeid, laboratorium, etc.) en een verlies aan productiecapaciteit. Het is dan ook van belang het filterrijpingsproces zo kort mogelijk te houden.

Het onderwerp van het onderzoek beschreven in dit proefschrift is de verwijdering van mangaan uit grondwater door middel van microbiologische en fysisch-chemische auto katalytische processen. De focus ligt hierbij op het filterrijpingsproces, waarbij de aandacht in bijzonder was gericht op het ontrafelen van de betrokken mechanismen en processen, met als doel de verkorting van de filterrijpingstijd.

Lange tijd is er van uitgegaan dat de verwijdering van mangaan uit grondwater, door middel van traditionele beluchting-filtratie, een fysisch-chemisch proces was, met een belangrijke rol voor het mangaanoxide Hausmannite. Later bleken ook microbiologische processen een rol te spelen. Ondanks het feit dat er veel onderzoek is gedaan naar de mangaanverwijdering, zijn de processen en mechanismen die hierbij betrokken zijn en hun onderlinge samenhang nog steeds niet volledig doorgrond. Doel van dit onderzoek was dan ook het vergroten van de kennis en de wijze waarop het filterrijpingsproces tot stand komt. Door de ontrafeling van het fenomeen filterrijping, kunnen oplosrichtingen om dit proces te verkorten geformuleerd worden.

Om het inzicht ten aanzien van mangaanverwijdering te vergroten, is bij de aanvang van dit onderzoek een inventarisatie uitgevoerd bij meer dan 100 grondwaterzuiveringsinstallaties, met name in Nederland, België en Duitsland. Op basis van deze inventarisatie was het mogelijk belangrijke parameters voor een succesvolle ontmanganing vast te stellen. Door statistische correlaties is aangetoond dat volledige ontmanganing, in aanwezigheid van ammonium en ijzer, mogelijk is wanneer voor een aantal parameters aan de volgende criteria is voldaan:

- NH_4^+ - verwijderingsefficiëntie : $> 85\%$
- Belading filter met ijzer, per filterrun : $< 2,7 \text{ kg Fe}/m^2.\text{FR}$
- pH van het filtraat : $> 7,1$
- filtratiesnelheid : $< 10,5 \text{ m/h}$
- Schijnbare verblijftijd : $> 11,5 \text{ min}$
- Zuurstofgehalte in filtraat : $\geq 1 \text{ mg/l}$

Een belangrijke stap in het ontrafelen van de processen en mechanismen die van belang zijn bij de start van het filterrijpingsproces is het vaststellen van het type mangaanoxide dat hierbij betrokken is. Ten behoeve van de karakterisatie en identificatie van mangaanoxide, aanwezig in filter media coatings, is gebruik gemaakt van een aantal analysetechnieken, te weten:

- Röntgen diffractie (XRD);
- Raman spectroscopie;
- Elektronenmicroscopie en "Energy Dispersive X-ray analysis" (SEM-EDX);
- Elektron paramagnetische resonantie (EPR).

Op basis van de combinatie van bovenstaande analysetechnieken is vastgesteld dat in alle onderzochte monsters van filter media coatings, een amorf mangaanoxide van het type Birnessite aanwezig was. Van Birnessite is bekend dat het een mangaanoxide is met uitstekende adsorptie eigenschappen voor diverse ionen, waaronder Mn^{2+}. Verder heeft Birnessite ook auto katalytisch oxidatieve eigenschappen. Beide eigenschappen maakt Birnessite uitermate geschikt voor een effectieve en efficiënte verwijdering van mangaan uit grondwater.

Een volgende stap in het onderzoek was het volgen van het rijpingsproces. Hiervoor is een proefinstallatie onderzoek uitgevoerd op productielocatie Grobbendonk van het Vlaamse waterbedrijf Pidpa. Gedurende het gehele rijpingsproces zijn monsters genomen van zowel het filtermateriaal als het spoelwater. Met behulp van de beschreven analysetechnieken is het gevormde mangaanoxide gekarakteriseerd en geïdentificeerd. Bovendien is het met de combinatie van SEM en EPR vastgesteld of Birnessite, biologisch of fysisch-chemisch is gevormd. Op basis van deze technieken kon worden vastgesteld dat de vorming van het mangaanoxide (Birnessite) op biologische wijze was gestart. Gedurende het rijpingsproces nam het aandeel fysisch-chemisch gevormd Birnessite toe. Na ca. 450 - 550 dagen bleek de Birnessite, aanwezig op het filter materiaal, volledig op fysisch-chemische wijze gevormd te zijn. Birnessite, aanwezig in het spoelwater, bleek gedurende het gehele filterrijpingsproces echter vooral op biologische wijze te zijn geproduceerd.

Tijdens het pilotonderzoek is ook de bacteriepopulatie, aanwezig in de mangaanverwijderingsfilters, onderzocht. Bij dit onderzoek is gebruik gemaakt van een aantal moleculaire analysetechnieken:

- "Next generation DNA-sequencing";
- qPCR;
- MALDI-TOF MS analyse.

Tijdens dit onderzoek is een duidelijke verschuiving van de bacteriepopulatie tijdens de start van het ontmanganingsproces aangetoond. In het filter dat als voedingswater diende voor het mangaanverwijderingsfilter waren met name bacteriën van het geslacht *Gallionella* aanwezig (>97% van de totale bacterie populatie). Bacteriën, aanwezig in het spoelwater van het mangaanverwijderingsfilter, kort nadat de ontmanganing volledig was, bestond nog voor "slechts" 12,4 % uit bacteriën van het geslacht *Gallionella*. Verder waren de bacterie geslachten: *Nitrospira* (25,7%) en *Pseudomonas* (14,3%) aanwezig. Ongeveer 47.6% van de bacteriepopulatie bestond uit kleine bacteriegroepen, die veelal niet volledig gekarakteriseerd zijn. Van *Nitrospira* is bekend dat het in staat is nitriet om te zetten in nitraat als onderdeel van de ammoniumoxidatie. De aanwezigheid van dit bacterie geslacht is dus verklaarbaar, omdat naast mangaan ook ammonium werd omgezet in dit filter. Van *Pseudomas* sp., en in het bijzonder *Pseudomonas putida*, is bekend dat het in staat is mangaan te oxideren.

Uit qPCR-analyses is echter gebleken dat de species *P. putida*, slechts in zeer geringe mate aanwezig was (< 0,01%). Met behulp van MALDI-TOF MS zijn een aantal nauw verwante *Pseudomonas* soorten aangetroffen, onder meer: *P. gessardii*, *P. grimontii* en *P. koreensis*. Het lijkt er dan ook op dat het bacteriegeslacht *Pseudomonas* betrokken is bij de start van het filterrijpingsproces. Tijdens dit onderzoek is het, onder gecontroleerde laboratoriumcondities, niet gelukt geïsoleerde *Pseudomonas* soorten Mn^{2+} te laten oxideren, dit in tegenstelling tot een laboratoriumstam van *P. putida*. Of mangaanoxidatie in de proefinstallatie van Grobbendonk een solitaire actie is van deze bacteriegeslacht, dan wel een gezamenlijk proces met een consortium van ook andere bacteriën, is niet vastgesteld tijdens dit onderzoek.

De effectiviteit van Birnessite in filter media met mangaanoxide coating (MOCS/MOCA) is vervolgens op praktijkschaal getest. Verse MOCA was in staat de rijpingstijd tot nul te reduceren. Droog MOCS was niet in staat de rijpingstijd te verkorten. Bovenstaande testen zijn echter uitgevoerd op twee verschillende locaties onder verschillende operationele condities. Bovendien was ook de kwaliteit van het grondwater op beide locaties verschillend.

Op basis van bovenstaande bevindingen zijn vervolgens, op proefinstallatieschaal, testen uitgevoerd met droog en vers gecoat filtermateriaal onder dezelfde condities. Tijdens dit onderzoek is nieuw filter materiaal (zand en antraciet) gebruikt als referentie. Uit deze experimenten is geconcludeerd dat er bij gebruik van nieuw zand en nieuw antraciet, sprake was van vergelijkbare filterrijpingstijden. Verder is vastgesteld dat bij toepassing van zowel vers MOCA als vers MOCS de filterrijpingstijd is gereduceerd tot nul. Droog gecoat filtermedia leidde slechts tot een tijdelijke adsorptie van Mn^{2+}.

Verder is tijdens dit onderzoek de invloed van de terugspoeling van een filter op het filterrijpingsproces onderzocht. De frequentie van het terugspoelen van een filter had een belangrijke, negatieve, invloed op de start van het filterrijpingsproces. Hiermee vervult de filterspoeling dus een sleutelrol en is daarmee een zogenaamde sleutelfactor bij de start van het ontmanganingsproces.

Gedurende het filterrijpingsproces neemt de invloed van de filterspoelingen af, ten gevolge van de ontwikkeling van een dikkere laag biomassa en/of auto katalytisch actief Birnessite.

Filterspoelingen hebben geen of slechts een marginale invloed op de mangaanverwijdering, wanneer gebruik gemaakt wordt van een laag vers MOCS of MOCA.

De terugspoelfrequentie van een filter is vooral afhankelijk van de hoeveelheid geoxideerd Fe^{2+} ($Fe(OH)_3$), die wordt afgevangen in het filter (de filterbelading met ijzer). Ook de belading van het filter met ijzer en de concentratie van Fe^{2+} in ruw water, kunnen hierdoor worden beschouwd als sleutelfactoren met betrekking tot de start van het ontmanganingsproces.

Resumerend kan worden gesteld dat in dit proefschrift nieuwe en belangrijke inzichten zijn gepresenteerd met betrekking tot het ontrafelen van processen en mechanismen die betrokken zijn bij (de start van) het filterrijpingsproces.

Aangetoond is dat Birnessite, en niet Hausmanite, een cruciale rol speelt bij ontmanganing. De vorming van Birnessite begint biologisch, maar gedurende het ontmanganingsproces blijkt de vorming van fysisch-chemisch auto katalytisch Birnessite dominant. Bij de biologische ontmanganing spelen naar alle waarschijnlijkheid meerdere bacteriesoorten een rol, waaronder bacteriën van het geslacht *Pseudomonas*.
Op basis van deze inzichten is het voor waterbedrijven mogelijk, bij de engineering en in de dagelijkse bedrijfsvoering, maatregelen te nemen om het filterrijpingsproces te optimaliseren. Deze maatregelen kunnen zijn gericht op het creëren van optimale condities voor mangaanoxiderende bacteriën, waarbij met

name de frequentie van de filterspoeling wordt beperkt (b.v. door het beperken van de belasting van het filter met ijzer). Verder kan gebruik gemaakt worden van vers gecoat filter media (MOCS/MOCA), waardoor volledige eliminatie van de filterrijpingstijd realiseerbaar is.

Contents

Figure: *Filter media for manganese removal (clockwise from top) ; MOCS, Virgin Sand, Extruded Activated Carbon, Glass, Anthracite, IOCS, Zeolite and Limestone (photo by J.H. Bruins, 2010)*

1 GENERAL INTRODUCTION

"Access to safe drinking water is essential to health, a basic human right and a component of effective policy for health protection"
(WHO, 2011)

1.1 Manganese in groundwater and groundwater treatment in The Netherlands

1.1.1 Manganese occurrence in groundwater

Groundwater is an important source for drinking water production in The Netherlands (Vewin, 2012), in Europe (EU, 2008) and world-wide (UNEP, 2008). Just like iron, manganese is a commonly occurring contaminant present in most groundwater (WHO, 2004).

In nature, manganese occurs as a compound, found in many types of rocks. It is usually found together with iron and silica, and is the 10th most abundant element in the earth's crust (ATSDR, 2008; IMnI, 2010). It is a constituent of more than 30 manganese oxide/hydroxide based minerals, playing an active role in the environmental geochemistry (Post, 1999). Manganese oxides are ubiquitous in soils and sediments, and because they are highly chemically reactive and strong scavengers of heavy metals, they exert considerable influences on the chemical behaviour of sediments, soils and associated aqueous systems. Manganese can exist in multiple oxidation states. The environmentally and biologically most important minerals are those containing Mn^{2+} or Mn^{4+} (USEPA, 2004). Due to the natural occurrence of manganese in sediments and soils, manganese is also present in associated aqueous systems with low redox potential and pH, such as anaerobic groundwater. The occurrence and thermodynamic stability of different manganese species in natural waters depends on conditions like redox, pH, temperature and oxygen concentration. The most abundant and stable manganese species in anaerobic groundwater (low pH and redox potential) is Mn^{2+} (Stumm and Morgan, 1996).

1.1.2 The relevance of manganese in drinking water and guideline values

Trace levels of manganese are essential for growth and development of humans, animals and plants. However, for health and aesthetic considerations the amount of manganese in drinking water should be limited to very low values.

Health concern associated with manganese presence in drinking water is mostly related to neurologic symptoms (Santamaria and Sulsky, 2010; Rodriguez *et al.*, 2013,). Manganese-induced clinical neurotoxicity is also associated with a motor dysfunction syndrome commonly referred to as Manganism (a Parkinson-like disorder). Dosages of 1 to 150 mg/kg of body weight per day ('short term exposure') and 1 to 2 mg/kg body weight per day ('long term exposure') (WHO, 2004) of oral or inhalation exposure are associated with increased manganese levels in tissue. That may lead to the development of these adverse neurological, reproductive, or respiratory diseases. However, an increasing number of studies report associations between neurologic symptoms and manganese exposure in infants and children. These findings, in combination with the questionable scientific background of data used in setting the current manganese WHO-guideline value for drinking water at 400 µg/L, warrant re-evaluation of the guideline (WHO, 2004; Ljung and Vather, 2007). Especially the manganese uptake through drinking water consumption by babies could be relatively high due to their low body weight, relatively high intake and relatively poor manganese excretion (Brown and Foos, 2009). In addition, the presence of manganese exacerbates the health problems caused by arsenic (Bunderson *et al.*, 2006, Wright *et al.*, 2006). Thus, the presence of manganese in drinking water is of particular risk in those areas that also contain elevated levels of arsenic. Therefore the WHO-health based guideline value for manganese in drinking water is 400 µg/L (WHO, 2011). In the US, the lifetime health advisory value is 300 µg/L and the EPA's Secondary Maximum Contaminant Level is 50 µg/L. The former WHO guideline value of 100 µg/L is still valid in several countries (*e.g.*, Serbia and Jordan) while the guideline value of the EU is 50 µg/L (EU, 1998). This is also the guideline value for manganese according to the Dutch Revised Drinking Water Decree (VROM, 2011). The internal standard for several Dutch and Belgian Water Supply Companies varies from 5 to 20 µg/L based on aesthetic considerations and practical problems in drinking water treatment.

1.1.3 Groundwater treatment in the Netherlands

In The Netherlands, 77 % of the groundwater sources, contain manganese (Mn^{2+}) in concentrations higher than the EU and Dutch (NL) guideline value of 50 µg/L, whereas approximately 14 % contains more Mn^{2+} than the health related WHO guideline value of 400 µg/L (Fig. 1.1).

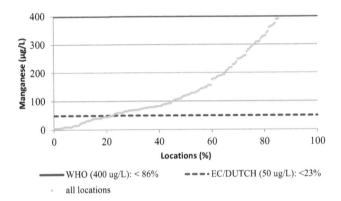

Figure 1.1: *overview of manganese concentrations in Dutch groundwater (2013), together with the WHO, EU and Dutch (NL) guideline values for manganese in drinking water (RIVM, 2013)*

In The Netherlands, manganese is removed from groundwater by aeration-rapid (sand) filtration (Fig. 1.2). Such a treatment process is easy to operate, cost effective and sustainable, as does require use of strong oxidants (*e.g.*, O_3, Cl_2, ClO_2, $KMnO_4$), with the associated risk of disinfection by-product formation and over or under dosing. Also this process does not require use of special adsorbents (*e.g.*, manganese green sand), which are applied in other countries (*e.g.*, USA) and which must be regenerated in time (Knocke *et al.*, 1991). However, the application of aeration-filtration in practice is not completely problem-free.

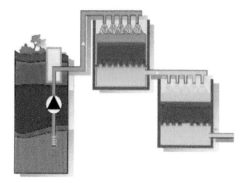

Figure 1.2*: Typical aeration-filtration groundwater treatment scheme with pre aeration, (down flow) first rapid filtration step (sand), second aeration and second rapid filtration step (sand) (source: Water Supply Company Groningen)*

1.2 Problem description regarding manganese removal in practice

In Western European countries such as The Netherlands, Germany and Belgium, problems caused by manganese are related to aesthetic considerations and practical problems in drinking water treatment, rather than health related problems. Aesthetic considerations are organoleptic properties like undesirable taste, staining of plumbing fixtures and laundry (WHO, 2004) and deposition of manganese oxides in distributions systems, causing black water incidents (Sly *et al.*, 1990). Practical problems in drinking water treatment associated with manganese presence, express themselves in blocking of filter nozzles and clogging of filters, valves and piping.

To remove manganese from groundwater, aeration followed by rapid sand filtration, is commonly applied. However, manganese removal by aeration-filtration is frequently associated with problems, such as:

- very long ripening periods (several months to more than one year) are required to achieve an effective removal with new filter media, leading to associated costs and reduced production capacity (Cools, 2010; Huysman, 2010; Krull, 2010);
- frequent manganese breakthrough of filters after some years of operation, resulting in filter media replacement. This phenomenon is again associated with long start-up and additional costs for filter media disposal and replacement (De Ridder, 2008);
- inefficient manganese removal from some groundwater types, *e.g.*, with a low pH (< 7) , and a high concentration of ammonia ($NH_4 > 2.5$ mg/L) (Gouzinis *et. al.*, 1988).

Particularly, the long ripening time of filter media is a major concern and is subject of this study. Due to the long ripening time, water companies have to waste large volumes of treated water, making this process less sustainable. In 2013, the Water Company Groningen had to waste more than 100,000 m^3 of pre treated water during start up of one filter with virgin filter media. In addition, costs associated with filter media ripening (man power, electricity, operational and analysis costs) are high. Furthermore, a filter in the process of ripening cannot be used for the production of drinking water. Decreasing filter ripening time for manganese removal is a serious concern for water supply companies. Several mechanisms are suggested to be involved in the startup of manganese removal filters with virgin media, and include the following:

- physical and chemical processes, *e.g.,* adsorption, oxidation, (co)precipitation;
- microbiological manganese oxidation and removal;
- combinations of the aforementioned processes.

Despite the fact that already a lot of research has been done on manganese removal, the controlling mechanisms, especially of the startup of filter media ripening, are not yet fully understood. Several research publications suggest that ripening time in manganese removal by conventional aeration-filtration could be lowered by introducing manganese and/or iron (hydro) oxide coated filter media to conventional rapid sand filters (Buamah, 2009; Buamah *et al.,* 2008, 2009a; Hu *et al.* 2004a,b; Islam *et al.,* 2010; Kim *et al.,* 2009; Olanczuk-Neyman *et al.,* 2000; Sahabi *et al.,* 2009; Stembal *et al.,* 2005; Tekerlekopoulou and Vayenas, 2008; Tiwari *et al.,* 2007). Results indicated that oxidation of manganese and iron, adsorbed on filter media during groundwater treatment play an important role in good long-term performance of manganese removal filters (Buamah *et al.,* 2009). Furthermore, results obtained by Buamah (2009) showed that water quality (*e.g.,* pH and HCO_3^-) also plays an important role in manganese removal by influencing the solubility of Mn^{2+} (Fig. 1.3).

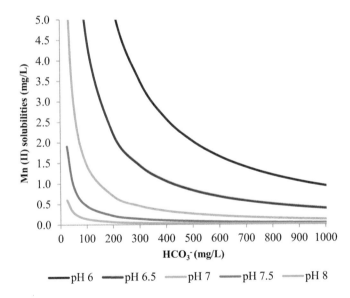

Figure 1.3: *solubility of Manganese (Mn^{2+}) as a function of pH and HCO_3^- concentration (adopted from Buamah, 2009)*

It is also known from literature that biology may play an important role with respect to manganese removal. Several researchers investigated the influence and the capability of different types of bacteria to oxidize manganese, amongst others *Leptothrix* sp. (Adams and Ghiorse, 1985; Barger et al., 2009; Boogerd and De Vrind, 1987; Burger *et al.,* 2008a; Burger *et al.,* 2008b; Corstjens *et al.,* 1997; El Gheriany *et al.* 2009; Hope and Bott, 2004; Tebo *et al.* 2004, 2005); *Pseudomonas* sp. (Barger *et al.,* 2009; Brouwers *et al.,* 1999; Caspi *et al.,*

1998; DePalma, 1993; Gounot, 1994; Tebo *et al.*, 2004, 2005; Villalobos *et al.*, 2003, 2006) and *Bacillus* sp. (Barger *et al.*, 2005, 2009; Brouwers *et al.*, 2000; Mann *et al.*, 1988; Tebo *et al.*, 2004, 2005). However, the presence of bacteria that are potentially able to oxidize manganese, is not a guarantee of substantial manganese oxidation and subsequent manganese removal in filters.

Consequently, additional research is still required. In particular, the type of MnO_x present in filter media coatings needs to be identified, how this manganese oxide coating is formed on virgin filter media and the role of biology in the process. Also the manganese adsorption capacity of manganese coated filter media, such as manganese oxide coated sand (MOCS) and manganese oxide coated anthracite (MOCA) has to be investigated, as well as their role in the adsorption and subsequent oxidation of manganese. Furthermore, knowledge of parameters playing an important role in manganese removal such as water quality, operational and design parameters is limited, and should be investigated in more detail if the process is to be optimized.

Therefore, additional information is required on:

- key factors required for efficient manganese removal in practice, in conventional aeration filtration ground water treatment plants (GWTPs);
- the effect of raw water quality parameters (*e.g.*, pH, Fe^{2+}, NH_4^+) on manganese removal;
- the type of MnO_x present in filter media coatings, like MOCS and MOCA;
- the mechanisms responsible for a rapid start up of virgin filter media ripening;
- the type of MnO_x formed initially on virgin filter media;
- the contribution of biological processes in the start up of manganese removal.

1.3 Aim and research objectives of this thesis

The main goal of this PhD research is to get a better understanding of the mechanisms involved in the ripening of virgin filter media used in manganese removal. Furthermore, this research will allow the development of an innovative manganese removal process that can shorten or completely eliminate ripening in new filters, as well as prolonging the lifetime of filter media.

In order to achieve the established goals, the following research objectives were defined:

1. Prepare an overview of manganese removal in selected aeration filtration groundwater treatment plants, including groundwater quality, process conditions applied and operational experiences.
2. Characterise and identify the manganese oxide(s) present in filter media coatings in conventional aeration-filtration groundwater treatment plants, with efficient manganese removal.
3. Provide a better understanding of the processes involved in ripening virgin filter media by performing ripening experiments at pilot scale, and propose measures to speed up filter media ripening and subsequent manganese removal.
4. Investigate the role of biological processes in manganese removal on laboratory, at pilot (batch) and full scale and propose beneficial conditions for manganese oxidizing bacteria.
5. Investigate the potential of filter media with a high manganese oxide content (*e.g.*, MOCS and MOCA) to accelerate filter media ripening in conventional aeration filtration filters in practice, and propose measures resulting in improved manganese removal.

6. Improve understanding of process and operational conditions applied in practice (*e.g.,* back washing, iron load) in conventional aeration-filtration systems and identify key factors to optimize manganese removal in practice.

1.4 Outline of this thesis

The introduction of this thesis (chapter 1) describes the problem statement as well as general information about groundwater treatment and the need for manganese removal, based on health, aesthetic and operational aspects. This chapter ends with the goal of this study, the research objectives and the thesis outline.

Chapter 2 deals with water quality and operational parameters that affect manganese removal in aeration - rapid sand filtration systems. Data from selected full-scale groundwater treatment plants in The Netherlands, Belgium, Germany, Jordan and Serbia were collected. The focus of the overview was to assess the effect of groundwater quality, process design and operational parameters on manganese removal efficiency in the first aeration-filtration stage of treatment plants. In the first filtration stage, manganese is removed together with iron and ammonia. Single and multiple statistical relations between manganese removal and water quality data, process design and operational parameters were conducted to define key factors responsible for manganese removal in practice.

Chapter 3 covers the characterization and identification of naturally formed manganese oxide coatings that can remove Mn^{2+} in conventional groundwater treatment plants (GWTPs). Characterization was carried out via X-ray diffraction (XRD), Scanning Electron Microscopy (SEM) and Raman spectroscopy analyses. Furthermore based on Electron Paramagnetic Resonance (EPR) analysis, the biological or physico-chemical origin of the manganese oxide present in the filter media coating was established.

Chapter 4 describes the ripening process of virgin media in a pilot under well defined and controlled conditions. This media ripening process was followed for more than 600 days. During this study, manganese oxide on filter sand and in the backwash water was investigated by the characterization and identification techniques mentioned in chapter 3. The origin of the formed manganese oxide at the start of filter media ripening and during prolonged filter media ripening was determined. Based on the evolution of manganese oxide formed during ripening, the mechanisms involved were explained.

Chapter 5 deals with the role of biology in filter ripening and subsequent manganese removal. In this study advanced molecular techniques, such as qPCR, DNA pyro sequencing and Malditof protein analyses were employed throughout the pilot research phase mentioned in chapter 4. This study also yields information on bacteria species present during ripening. Furthermore in this chapter the conditions beneficial for biological manganese removal are described.

Chapter 6 describes the results of filter ripening in practice and the influence of employing coated filter media. The filter media ripening with and without coated media were investigated at two full scale groundwater treatment plants. Different virgin filter media (anthracite and sand) were employed at each location and Manganese oxide coated sand (MOCS) was added in one location and manganese oxide coated anthracite (MOCA) at the other location. The aim of this study was to get information about the adsorptive capacity of the two filter media coatings and their effect on filter media ripening in practice.

Chapter 7 recommends some key factors to speed up filter media ripening in practice. Therefore, the benefits of using coatings and the differences between the use of MOCS and MOCA in practice (chapter 6) are further investigated in a pilot research (at GWTP Grobbendonk). In this pilot research, the focus was to investigate the influence of: (1) back washing (or filter loading with particles), (2) the differences between selected virgin media (anthracite and sand) on filter ripening, (3) differences in the ripening times between virgin filter media and MOCA and MOCS and (4) the use of dry and fresh coated filter media. Based on the outcome and results of this study, a strategy to optimize ripening is outlined.

Finally, chapter 8 summarizes the main results, conclusions and limitations of this research. Also recommendations for practical use and suggestions for follow up research are given.

1.5 References

Adams, L.F., Ghiorse, W.C.1985. Influence of Manganese on Growth of a Sheathless Strain of *Leptothrix discophora*. Aplied and environmental Microbiology, 49 (3), 556-562.

ATSDR. 2008. (draft, update 2000) Toxicological profile for manganese, U.S. Department of Health and Human Services, Agency for Toxic Substances and Disease Registry, Atlanta.

Barger J.R., Tebo, B.M., Bergmann, U., Webb, S.M., Glatzel, P., Chiu, V.Q., Villalobos, M. 2005. Biotic and abiotic products of Mn(II) oxidation by spores of the marine Bacillus sp. strain SG-1. American Mineralogist, 90, 143-154.

Barger, J.R., Fuller. C.C., Marcu. M.A., Brearly A., Perez De la Rosa M., Webb S.M., Caldwell W.A. 2009. Structural characterization of terrestrial microbial Mn oxides from Pinal Ckeek, AZ. Ceochimica et Cosmochimica Acta 73, 889-910.

Boogerd, F.C., De Vrind J.P.M. 1987. Manganese oxidation by *Leptothrix discophora*, Journal of bacteriology, 489-494.

Brouwers, G.J., Vrind de J.P.M, Corstjens P.L.A.M., Cornelis P, Baysse C, Vrind de Jong E.W. 1999. CumA, aGene Encoding a multicopper oxidase, is involved in Mn^{2+}oxidation in *Pseudomonas putida* GB-1. Applied and environmental microbiology , 65 (4), 1762-1768.

Brouwers G.J., Vijgenboom E., Corstjens P.L.A.M., de Vrind J.P.M., de Vrind-de Jong E.W. 2000. Bacterial Mn^{2+} oxidizing systems and multicopper oxidases: an overview of Mechanisms and Functions, Geomicrobiology Journal, 17, 1-24.

Brown M.T., Foos B. 2009. Assessing Children's exposure and risks to drinking water contaminants: A Manganese case study, Human and Ecological Risk Assessment, 15, 923-947.

Buamah, R., Petrusevski, B., Schippers, J.C. 2008. Adsorptive removal of manganese (II) from the aqueous phase using iron oxide coated sand. Journal of Water supply: Research and Technology-AQUA 57.1, 1-11.

Buamah, R. 2009. Adsorptive removal of manganese, arsenic and iron from groundwater. PhD-thesis, UNESCO-IHE Delft / Wageningen University, The Netherlands.

Buamah, R., Petrusevski, B., de Ridder, D., van de Watering and Schippers, J.C., 2009a. Manganese removal in groundwater treatment: practice, problems and probable solutions. Journal of Water Science and Technology: Water Supply 9.1, 89-98.

Bunderson M., Pereira F., Schneider M.C., Shaw P.K., Coffin J.D., Beall H.D. 2006. Manganese enhances peroxynitrite and leukotriene E4 formation in bovine aortic endothelial cells exposed to arsenic, Cardiovasc. Toxicol., 6 (1), 15-23.

Burger, M.S., Krentz, C.A., Mercer, S.S., Gagnon, G.A. 2008a. Manganese removal and occurrence of manganese oxidizing bacteria in full-scale biofilters. Journal of Water Supply: Research and technology-AQUA 57.5, 351-357.

Burger, M.S., Mercer, S.S., Shupe, G.D., Gagnon, G.A. 2008b. Manganese removal during bench-scale biofiltration, Water Research 42, 4733-4742.

Caspi, R., Tebo, B.M., Haygood, M.G. (1998). C-Type Cytochromes and Manganese Oxidation in *Pseudomonas putida* MnB1. Applied and environmental Microbiology, 64.10, 3549-3555.

Cools B. 2010. De Watergroep, personal communication, Belgium.

Corstjens, P.L.A.M., de Vrind, J.P.M., Goosen, T., de Vrind-de Jong, E.W. 1997. Identification and molecular analysis of the Leptothrix discophora SS-1 mofA gene, a gene putatively encoding a manganese-oxidizing protein with copper domains. Geomicrobiology Journal, 14 (2), 91-108.

DePalma, S.R. 1993. Manganese oxidation by *Pseudomonas putida*, PhD-thesis, Harvard University, Cambridge, Massachusetts, USA.

El Gheriany I.A., Bocioaga B., Anthony Hay A.D., Ghiorse W.C., Shuler M.L., Lion L.W. 2009. Iron Requirement for Mn(II) Oxidation by *Leptothrix discophora* SS-1, Applied and Environmental Microbiology, 75 (5),1229-1235.

EU. 1998. *Drinking Water Directive*, Council directive 98/83/EC.

EU. 2008. Groundwater protection in Europe - The new groundwater directive, European commission.

Gounot A-M. 1994. Microbial oxidation and reduction of manganese: Consequences in groundwater and applications, FEMS Microbiology reviews 14, 339-350.

Gouzinis, A., Kosmidis N., Vayenas D.V., Lyberatos G. 1988. Removal of Mn and simultaneous removal of NH_3, Fe and Mn from potable water using a trickling filter, Wat. Res. 32 (8), 2442-2450.

Hope C.K., Bott T.R. 2004. Laboratory modelling of manganese biofiltration using biofilms of *Leptothrix doscophora*, Water Research 38, 1853-1861.

Hu, P-Y., Hsieh, Y.-H., Chen, J-C., Chang, C-Y. 2004a. Adsorption of divalent manganese ion on manganese-coated sand. Journal of Waters Supply: Research and Technology-AQUA 53.3, 151-158.

Hu, P-Y., Hsieh, Y.-H., Chen, J-C., Chang, C-Y. 2004b. Characteristics of manganese-coated sand using SEM and EDAX analysis. Journal of Colloid and interface science, 272, 308-313.

Huysman K. 2010. fPidpa Department of Process Technology and Water Quality, personal communication, Belgium.

IMnI. 2010. The international Manganese Institute, Paris, France, http://www.manganese.org.

Islam A.A., Goodwill, J.E., Bouchard, R., Tobiasen, J.E., Knocke, W.R. 2010. Characterization of filter media $MnO_2(s)$ surfaces and Mn removal capability. Journal AWWA, 102 (9), 71-83.

Kim, W.G., Kim, S.J., Lee, S.M., Tiwari, D. 2009. Removal characteristics of manganese-coated solid samples for Mn(II). Desalination and water treatment 4, 218-223.

Knocke, W.R., Van Benschoten, J.E., Kearny, M.J., Soborski, A.W., Reckhow, D.A. 1991.

Kinetics of Manganese and Iron oxidation by Potassium Permanganate and Chlorine dioxide, Journal of AWWA, June 1991, 80-87.

Krull J., Stadtwerke Emden (SWE). 2010 Personal communication, Germany.

Ljung K., Vahter M. 2007. Time to Re-evaluate the guideline value for manganese in Drinking water ?, Environmental Health Perspectives, 115 (11), 1533-1538.

Mann, S., Sparks. N.H.C., Scoot G.H.E., Vrind- De Jong E.W. 1988. Oxidation of Manganese and formation of Mn_3O_4 (Hausmannite) by spore Coats of a marine Bacillus sp. Applied and environmental microbiology, 54 (8), 2140-2143.

Olanczuk-Neyman, K., Bray, R. 2000. The role of Physico-Chemical and Biological Processes in Manganese and Ammonia Nitrogen Removal from Groundwater. Polish Journal of Environmental studies, 9 (2), 91-96.

Post, J. E. 1999. Manganese oxide minerals: Crystal structures and economic and environmental significance, Proceedings of the National Academy of Sciences USA, 96, 447-3454.

Ridder de D. 2008. Opstart mangaanverwijdering in snelfilters, BTO 2008.014 (in Dutch).

RIVM - National Institute for Public Health and the Environment. 2013. Information groundwater composition in The Netherlands 2012, provided by email, Mr. Dik.

Rodríguez-Barranco M., Lacasaña M., Aguilar-Garduño C., Alguacil J., Gil F., González-Alzaga B., Rojas-García A. 2013. Association of arsenic, cadmium and manganese exposure with neurodevelopment and behavioral disorders in children: A systematic review and meta-analysis. Science of the Total Environment 454-455, 562-577.

Sahabi, D.M., Takeda M., Suzuki I., Koizumi J-I. 2009. Removal of Mn^{2+} from water by "aged" biofilter media: The role of catalytic oxides layers. Journal of Bioscience and Bioengineering, 107 (2), 151-157.

Santamaria, A. B., Sulsky S.I. 2010. Risk assessment of an essential element: Manganese, Journal of Toxicology and Environmental Health, part A, 73, 128-155.

Sly L.I., Hodgkinson M.C., Vulapa Arunpairojana. 1990. Deposition of manganese in a drinking water distribution system. Applied and environmental biotechnology, 56 (3), 628-639.

Stembal, T., Markic, M., Ribicic, N., Briski, F., Sipos, L. 2005. Removal of ammonia, iron and manganese from ground waters of Northern Croatia – pilot plant studies, Process Biochemistry 40, 327-335.

Stumm, W. and Morgan, J.J. 1996. Aquatic Chemistry, chemical equilibria and rates, 3rd ed. Wiley, New York.

Tebo, B.M., Marger J.R., Clement B.G., Dick G.J., Murray K.J., Parker. D., Verity R., Webb S.M. 2004. Biogenic Manganese oxides: Properties and mechanisms of formation. Annu. Rev. Earth Planet Sci., 32, 287-328.

Tebo, B.M., Johnson H.A., McCarthy J.K., Templeton A.S. 2005. Geomicrobiology of manganese (II) Oxidation. TRENDS in Microbiology, 13 (9), 421-428.

Tekerlekopoulou, A.G. & Vayenas D.V. 2008. Simultaneous biological removal of ammonia, Iron and manganese from potable water using a trickling filter. Biochemical Engineering Journal 39, 215-220.

Tiwari, D., Yu, M.R., Kim, M.N., Lee, S.M., Kwon, O.H., Choi, K.M., Lim, G.J., Yang, J.K. 2007. Potential application of manganese coated sand in the removal of Mn (II) from aqueous solutions. Water Science & Technology, 56 (7), 153-160.

UNEP. 2008. Vital water graphics, http://www.unep.org/dewa/vitalwater/index.html.

USEPA. 2004. Drinking water Health Advisory for Manganese.

VEWIN. 2012 Drinkwaterstatistiek 2010, de watercyclus van bron tot tap (in Dutch).

Villalobos, M., Toner, B., Bargar, J., Sposito, G. 2003. Characterization of the manganese oxide produced by *Pseudomonas putida* strain MnB1. Geochimica et Cosmochimica Acta, 67 (14), 2649-2662.

Villalobos, M., Lanson, B., Manceau, A., Toner, B., Sposito, G. 2006. Structural model for the biogenic Mn oxide produced by *Pseudomonas putida*. American Mineralogist, 91, 489-502.

VROM. 2011. Ministry of Housing, Spatial Planning and the Environment, *Waterleidingbesluit* 2011 (in Dutch).

WHO. 2004. Manganese in Drinking water – Background document for development of WHO Guidelines for drinking water quality, Geneva, WHO/SDE/WSH/03.04/104.

WHO. 2011. Guidelines for drinking water quality – 4th edition. WHO Press, World Health Organization, 20 Avenue Appia, 1211 Geneva 27, Switzerland.

Wright R.O., Amarasiriwardena C., Alan D. Woolf A.D., Jim R., Bellinger D.C. 2006. Neuropsychological correlates of hair arsenic, manganese, and cadmium levels in school-age children residing near a hazardous waste site, NeuroToxicology, 27 (2), 210-216.

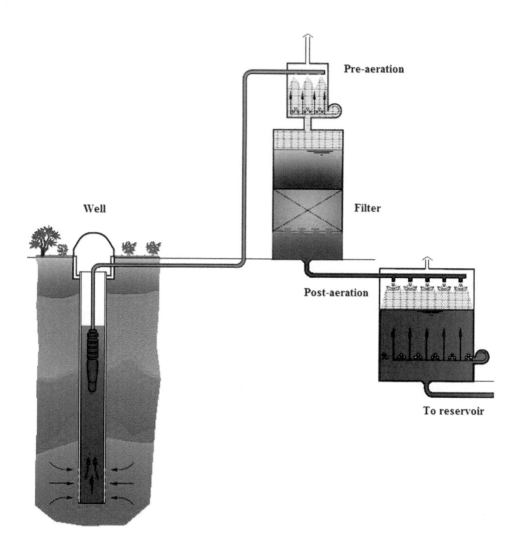

Figure: *Typical aeration-filtration groundwater treatment scheme (GWTP De Punt) with pre aeration, (down flow) single rapid filtration step (sand) and post aeration (source: Water Supply Company Groningen)*

2 ASSESSMENT OF MANGANESE REMOVAL FROM OVER 100 GROUNDWATER TREATMENT PLANTS

Main part of this chapter was published as:
*Jantinus H. Bruins, Dirk Vries, Branislav Petrusevski, Yness M. Slokar, Maria D. Kennedy (2014).
Assessment of manganese removal from over 100 groundwater treatment plants. Journal of Water
Supply: Research and Technology – AQUA 63.4:268 - 280*

2.1 Abstract

The aim of this study was to make an inventory of water quality and operational parameters which could affect manganese removal through aeration and rapid sand filtration and to establish correlations between these parameters and manganese removal efficiency. The focus of the overview was on manganese removal efficiency in the first aeration-filtration stage of conventional groundwater treatment plants. Data from selected full-scale groundwater treatment plants have been collected, and univariate and multivariate statistical analysis were conducted. Multivariate statistics indicated that multiple parameters including NH_4^+ removal efficiency, iron loading per filter run and pH of filtrate play a significant role in manganese removal, while other parameters (oxygen concentration in filtrate, filtration rate and empty bed contact time (EBCT)) were found to be of the secondary importance. Univariate statistical assessment of the data suggests that very effective manganese removal can be achieved when all of the following conditions are met: NH_4^+ removal efficiency > 85%, pH of filtrate > 7.1, iron loading per filter run < 2.7 kg Fe/m^2, oxygen concentration in filtrate > 1 mg O_2/L, filtration rate < 10.5 m/h and EBCT > 11.5 min.

Keywords: Aeration- filtration, conventional groundwater treatment plants, groundwater quality, inventory, manganese removal efficiency, operational parameters

2.2 Introduction

Groundwater is the predominant global source of water for drinking water production (UNEP 2000). In addition to impurities such as iron, ammonia and methane, groundwater frequently contains elevated levels of dissolved manganese, which need to be lowered for both health and aesthetic reasons. Manganese is a naturally occurring metal that is a constituent of more than 30 manganese oxide/hydroxide minerals, which occur in a wide variety of geological settings (Post, 1999). The occurrence and thermodynamic stability of different manganese species in groundwater's is dependent on different conditions like reduction-oxidation (redox) potential (E_h) or electron activity ($p\varepsilon$) and pH (Fig. 2.1).

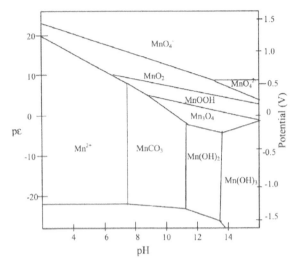

Figure 2.1: *Electron Activity (pε) and redox potential (E$_h$) – pH diagram for aqueous manganese (Stumm and Morgan, 1996)*

Due to the natural occurrence of manganese in sediments and soils, dissolved manganese (Mn^{2+}) can also be present in associated aqueous systems like groundwater, because of the low redox potential of anaerobic groundwater and the relative low pH. From Fig. 2.1 it can be seen that in a reducing environment and relatively low pH (conditions found in anaerobic groundwater), the most abundant and stable manganese species is Mn^{2+}.

The removal of manganese from groundwater is commonly achieved through aeration-rapid sand filtration. A typical example of a groundwater treatment plant (single filtration step with pre- and post-aeration) is shown in Fig. 2.2. An aeration followed by a single filtration step is often applied, but in some cases, *e.g.,* groundwater with elevated levels of ammonia, iron and/or manganese, two filtration steps are required. In those cases, a second aeration step is frequently included to correct the pH and increase the oxygen concentration in the (treated) water. Physical-chemical and/or biological mechanisms were reported to be responsible for manganese removal (Vandenabeele *et al.,* 1995; Graveland & Heertjes, 1975). Despite of extensive research carried out on manganese removal through aeration-rapid sand filtration, the mechanisms controlling the process are still not fully understood.

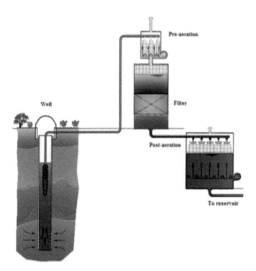

Figure 2.2: *Typical conventional groundwater treatment scheme with pre aeration, (down flow) single rapid filtration step and post aeration (source: Water Supply Company Groningen)*

Conventional groundwater treatment is simple and very cost effective, but, performance of an individual plant depends amongst other factors on individual approach to various technological issues related to a given groundwater. Advantages of manganese removal on conventional aeration-rapid sand filtration could be seriously hampered by:

(a) long ripening periods taking several weeks to more than a year to achieve effective manganese removal with virgin filter media (Huysman, 2010;).

(b) manganese breakthrough after some years of operation, introducing the need for filter media replacement, associated with a long start-up period and additional costs for filter media disposal and replacement (Buamah *et al.,* 2008). As an example in Fig. 2.3 the manganese breakthrough is expressed after more than 12 years of proper operating and treating the same raw water quality.

16

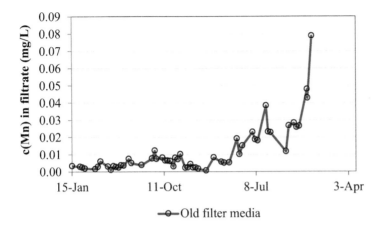

Figure 2.3: An example of manganese breakthrough in a Dutch groundwater treatment plant based on aeration-rapid sand filtration (source: Water Supply Company Groningen).

(c) insufficient manganese removal for some types of groundwater (*e.g.,* groundwater containing high concentration of Natural Organic Matter (NOM), that could cause complexation of Mn^{2+}).

Several researchers reported insufficient manganese removal due to water quality matrix and inappropriate process design and operational parameters. Amongst others, high concentrations of NH_4^+ (> 2 mg/L) can have a negative influence on manganese removal (Gouzinis *et al.,* 1998). In addition a low pH also negatively affect manganese adsorption (Buamah, 2009) and manganese oxidation (Stumm & Morgan, 1996). Without a catalyst, homogeneous manganese oxidation and subsequently removal by filtration is only achieved when pH is at least 8.6 (Graveland, 1971). Finally, dissolved iron present in groundwater can affect manganese removal, because it can compete with Mn^{2+} for adsorption sites (Hu *et al.,* 2004a, b). Therefore the main focus of this study was to improve the understanding of manganese removal in conventional groundwater treatment plants based on aeration and (down flow) rapid sand filtration. Specifically effect of groundwater quality, process design and operational parameters on manganese removal was assessed.

2.3 Materials and Methods

Data presented in this paper are based on information gathered from over 100 groundwater treatment plants (GWTPs) from The Netherlands (65), Belgium (34), Germany (6), Jordan (1) and Serbia (1). The data was mainly collected through questionnaire, or by visiting selected plants and interviewing key employees that have extensive operational experience. The focus of the inventory was on the plants that have a (down flow) single aeration- filtration treatment (with simultaneously removal of manganese, iron and ammonia) or on the 1st filtration step for plants that have multiple aeration-filtration steps. For GWTPs without a single aeration-filtration but with multiple aeration-filtration steps, questions were aimed at the first filtration step. Collected (water quality and process design) parameters where based on their importance for manganese removal, as suggested in literature (Buamah, 2009; Graveland, 1971; Gouzinis, 1998) but also depending on data availability (*e.g.,* redox potential is known to be of importance, but is seldom measured on regular base). Most of these plants included in the inventory have silica as filter media, in some cases in combination with anthracite as double media filter. Plants run for at least one year, so containing "aged" (bio)layers (*e.g.,* Fe and Mn oxides). The following data was collected or calculated.

Parameters with respect to the quality of anaerobic groundwater:

- concentration of ammonia, iron, manganese, hydrogen carbonate, calcium, phosphate, silica, methane, pH, dissolved or total organic carbon (DOC/TOC).

Parameters with respect to the water quality after the 1st filtration step:

- concentration of ammonia, iron, manganese, hydrogen carbonate, calcium, oxygen, pH;
- removal efficiency of manganese, ammonia and iron.

Parameters with respect to applied process design:

- iron and manganese loading (kg Fe and Mn/m^2/filter run);
- backwash (BW) criteria (head loss, time, volume);
- volume of filtrate produced between two consecutive backwashing cycles (m^3);
- filtration rate $(m^3/m^2/h)$;
- empty bed contact time EBCT (minutes) (t_{EB} in PCA analysis, Fig. 2.4);
- filter area (m^2);
- filter bed depth (m) (h_{FB} in PCA analysis, Fig. 2.4);
- flow (m^3/h);
- filter configuration (gravity/pressure), and type of filtration ('dry' or 'wet', with water level below or above filter media, respectively)

Selected raw and treated water quality and plant design parameters were correlated to the performance of full-scale plants in terms of Mn^{2+} removal. The focus was on the first filtration stage, in which Mn^{2+} removal was often combined with the removal of ammonia, iron and sometimes methane. Principal component analysis (PCA) and single correlation analysis of Mn^{2+} removal efficiency as a function of selected water quality parameters and process design parameters were conducted, to determine the parameters of importance. For PCA, data were standardized by subtracting the data values by the process variable means, and dividing the resulting data values by the standard deviation of the corresponding variable. To check whether the data set is useful for PCA, Kaiser-Meyer-Olkin (KMO) (Cerny *et al.,* 1977) and the Bartlett's tests (Bartlett, 1937) were performed. The KMO statistic assesses whether there is an underlying (latent) structure in the data. Low values of the KMO statistic, *i.e.,* smaller than an index of 0.6, indicate that the correlations between pairs of variables cannot be explained by other variables, and that PCA or factor analysis may not be appropriate. The Bartlett's test (Bartlett, 1937) is used to evaluate the homoscedasticity (equal variances across samples) of the data. A limit of 0.30 for the absolute value of the coefficient loading of a principal component is chosen to distinguish between moderate to high and low correlation.

2.4 Results and Discussion

2.4.1 Multivariate statistics and univariate correlations

Table 2.1 gives an overview of selected design and water quality parameters and correlation coefficients against manganese removal efficiency by a univariate and multivariate statistical method (PCA) .The KMO measure of sampling adequacy shows that the data set yields a degree of common variance of 0.65, which can be considered as sufficient for PCA analysis. Furthermore, Bartlett's test for dimensionality reveals that all the variables are necessary to explain non-random variations, in other words, they are correlated.

The first two components of the PCA are shown on the left of Fig. 2.4, while the number of components that account for the explained variance is shown on the right. In general, vectors pointing in the same direction mean that the associated factors are (positively) correlated, while if pointing away from each other, an inversely proportional relation exists. Additionally, loading vectors at right angles to each other indicate either a negligible or no interdependence. The variance of the variables in the data set is by definition of PCA covered by a summation of the principal components together with their loading coefficients. The part of variance that is covered by principal components, is called the explained variance. The first component explains the largest part of the variance in the data.

Table 2.1: Linear correlation coefficients and PCA loading coefficients between manganese removal efficiency and selected water quality and process parameters (PCA parameters with a value above ±0.30 are shown in bold face). PCA has been performed on standardized data.

Parameter /concentration			Correlation with Mn removal	PCA (N=34, rank of data set =17)	
	Parameter	Unit		1st component	2nd component
Manganese removal		%		**0.337**	-0.042
Process conditions	Filtration rate	m/h	-0.5046	-0.218	-0.248
	Bed height	m	-0.2859	-0.210	-0.265
	EBCT	min	0.3349	0.167	0.130
	Fe loading	kg/m^2	**-0.6861**	**-0.343**	-0.014
Raw water	Mn^{2+}	mg/L	-0.0362	-0.093	**-0.497**
	Fe^{2+}	mg/L	-0.5527	**-0.316**	0.032
	NH$_4^+$	mg/L	0.0354	0.157	0.315
	HCO$_3^-$	mg/L	0.3167	0.270	0.144
	Ca^{2+}	mg/L	0.3159	0.262	0.026
	PO$_4^{3-}$	mg/L	-0.2469	-0.204	**0.364**
	SiO$_2$	mg/L	-0.3732	0.112	0.013
	DOC/TOC	mg/L	0.0483	0.154	-0.126
Filtrate	NH$_4^+$	mg/L	-0.2018	-0.268	**0.312**
	NH$_4^+$ removal	%	**0.7618**	**0.331**	-0.090
	pH	-	**0.6164**	**0.333**	-0.045
	O$_2$	mg/L	0.0120	0.063	**-0.474**

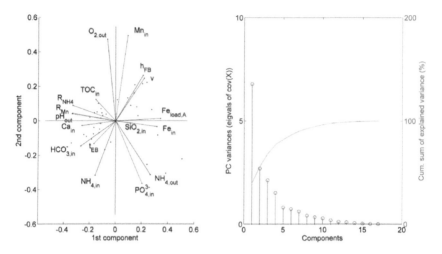

Figure 2.4: *PCA loadings shown as vectors in the principal component space (left) and number of components that account for the explained variance of the total data set (right).*

Based on both PCA and single correlation analysis (Table 2.1 and Fig. 2.4), NH_4^+ removal efficiency, iron loading per filter run, and pH strongly correlate with manganese removal efficiency. In contrast to NH_4^+ removal efficiency and pH, the iron loading (and often also the iron concentration in raw water) was found to be inversely proportional to manganese and NH_4^+ removal. Concentrations of dissolved calcium and hydrogen carbonate in raw water appear to have low correlation with manganese removal in single correlation studies, whereas PCA shows that these parameters could have some influence, as shown in the loadings of the first component. Since the solubility of compounds is directly dependent on pH, it was expected that calcium and hydrogen carbonate will be (cor)related with pH, thus influencing the loadings of the first component. Furthermore, the PCA analysis showed certain degree of correlation between all water quality and process parameters and manganese removal included in the inventory. This confirmed that Manganese removal in aeration-filtration treatment process is very complex and influenced by a large array of parameters. From the right hand side of Fig. 2.4 it follows that the second principal component explains another 15% of the variability in the data, while the third component covers an additional part of the variance, resulting in approximately 70% coverage of data variance when three components are used. Hence, the highest PCA loading coefficients of the second component (Table 2.1) indicate that, the influent manganese concentration, oxygen and phosphate might also play a role in manganese removal. A possible explanation for the negative (cor)relation between phosphate and manganese removal might be similar to that for the effect of phosphate on iron removal; heterogeneous Fe^{2+} oxidation (rate) decreases with increasing phosphate concentration. The controlling mechanisms are not fully understood yet, but the oxidation rate may be influenced by colloid formation and their mobility, the charge of filter media, and pH of (raw) water (O'Melia and Crapps, 1964; Wolthoorn *et al.*, 2004). An additional explanation could be that phosphate competes with Mn^{2+} for the same adsorption sites as observed for Fe^{3+} adsorption on iron hydro-oxide surfaces. The PCA further reveals that the variability in data cannot easily be explained by only 2 factors, i.e. at least 4 components are needed to account for approximately 75% of the explained variance (Fig. 2.4). Although the single correlation analysis does not confirm the effect of oxygen, organic matter TOC/DOC) and phosphate, it is known from practice that these parameters may also have effect on manganese removal efficiency. PCA shows that there is indeed an effect, although only significant in the second principal component for dissolved oxygen (Table 2.1), or even in the third for organic matter. It is

20

important to keep in mind that PCA informs how many parameters are necessary to explain the data and which variables are most importantly related (to manganese removal), while the following assessment is based primarily on evaluating the correlation between two variables and analysing the 'outliers' by inspection of the other variables.

In the following sections, the parameters with a predominant effect on manganese removal efficiency are discussed in more detail.

2.4.2 NH₄⁺ removal efficiency

Fig. 2.5 shows the correlation between manganese and NH_4^+ removal efficiencies for the treatment plant included in the inventory. Analysis of the data set (Fig. 2.5) suggests that for a majority of GWTPs, complete manganese removal is not achieved when NH_4^+ removal efficiency is less than 85%. A few GWTPs still did not achieve complete manganese removal even when NH_4^+ removal was very effective. Closer inspection reveals that each of the outliers had at least one parameter with values that were unfavourable for effective manganese removal, *i.e.,* pH in filtrate <7.1 or the filtration rate >10.5 m/h and/or EBCT <11.5 min (both parameters discussed further in this study).

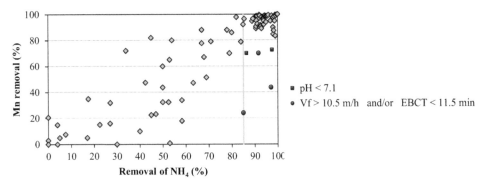

Figure 2.5: *Correlation between manganese and NH₄⁺ removal efficiencies for full-scale groundwater treatment plants based on aeration filtration.*

The strong correlation between NH_4^+ and manganese removal, is also observed in the practice. Very similar trends for NH_4^+ and manganese removal are specifically observed during the ripening period of new filter media (Fig. 2.6A). In addition, results from the full scale rapid sand filter suggest that complete manganese removal occurs only at a certain depth of a filter, when NH_4^+ removal is almost complete or NH_4^+ is not present anymore (Fig.2.6B).

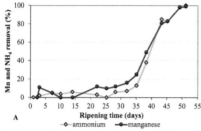 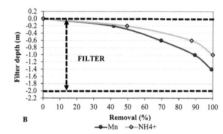

Figure 2.6: *Removal efficiency of NH$_4^+$ and manganese during the ripening period of new filter media of a full-scale groundwater treatment plant (water supply company Groningen, The Netherlands).* **A:** *Manganese and NH$_4^+$ removal efficiency in a single aeration-rapid sand filtration as a function of the ripening period.* **B:** *removal efficiency of manganese and NH$_4^+$ over the depth of a ripened filter.*

Several papers confirm that NH$_4^+$ removal and manganese removal are related. Scherer and Wichmann (2000) and Flemming *et al.* (2004), suggested that a negative effect of NH$_4^+$ on manganese removal is a consequence of poor Mn^{2+} oxidation due to the low redox potential of the groundwater in the presence of NH$_4^+$. As a result, high concentrations of NH$_4^+$ (>2 mg/L) in (raw) groundwater hinder efficient and complete manganese removal in conventional one-stage aeration-filtration systems Gouzinis *et al.* (1998). The observed correlation between NH$_4^+$ and manganese removal efficiencies may be also related to biological processes, *i.e.,* the oxidation of NH$_4^+$ by nitrifying bacteria. Several mechanisms may contribute:

- oxidation of Mn^{2+} by nitrifying bacteria (*e.g., Nitrosomonas* sp.), although from an energetic point of view this is unlikely (Vandenabeele *et al.*, 1995);
- Adsorption of positively charged dissolved Mn^{2+} on negatively charged organic compounds, *e.g.,* Extracellular Polymeric Substances (EPS) excreted by nitrifying bacteria (Vandenabeele *et al.*, 1995);
- Local reduction of the zero point of charge (pH$_{PZC}$) or Zeta potential of the filter media to more negative, due to a pH decrease on micro scale, induced by the nitrification reaction; more negatively charged filter media surface will considerably increase the attraction of positively charged Mn^{2+}.
- Increase of the redox potential due to the nitrification reactions resulting in enhanced Mn^{2+} oxidation.

However additional research is required to further clarify mechanisms responsible for the effect of NH$_4^+$ (removal efficiency) on manganese removal.

2.4.3 Effect of iron loading

Iron loading, defined as the amount of iron removed in a filter during one filter run, appears to negatively affect manganese removal (Table 2.1 and Fig. 2.7).

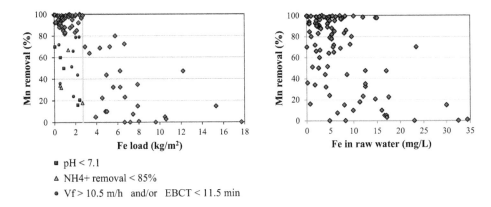

- ■ pH < 7.1
- △ NH4+ removal < 85%
- ● Vf > 10.5 m/h and/or EBCT < 11.5 min

Figure 2.7: *Manganese removal efficiency as a function of the iron loading per filter run (kg Fe/m²) and the iron concentration (mg/L) in raw groundwater*

From Fig. 2.7, it seems that complete manganese removal cannot be achieved with an iron loading above approximately 2.7 kg Fe/m² per filter run. Although iron oxides can adsorb Mn^{2+}, and can consequently also act as a catalyst for manganese removal (Buamah, 2009), iron (hydr)oxide can also cover active sites on filter media available for Mn^{2+} adsorption (Buamah *et al.*, 2009). In addition Fe^{2+} ions can compete with Mn^{2+} ions for adsorption sites (Hu *et al.*, 2004a,b). As a consequence, better manganese removal is typically achieved when either iron is removed prior to manganese removal, or when iron loading per filter run is lower. The latter can be achieved by reducing the amount of water filtered between two backwash cycles by *e.g.*, reducing the filter run length. Data from the conducted inventory also suggest that complete manganese removal in a single aeration-filtration treatment can be achieved also for groundwater with very high iron content (up to approximately 15 mg/L) when iron loading per filter run is limited to < 2.7 kg/m² (Fig. 2.7). Iron loading can also be influenced by the prevailing iron removal mechanism. The achieved iron loading is typically much lower when iron is removed through oxidation-floc formation in comparison to adsorptive and/or biological iron removal. Iron removal through oxidation and floc formation is characterised by accumulation of iron (hydr)oxide flocs in the upper part of the filter bed or in the anthracite part of a dual media filters, and associated faster head loss increase, and the need for frequent backwashing, specifically when fine sand size fraction is used. Consequently, iron removal predominantly through floc formation results in relatively lower iron loading and has potentially a larger adsorption/oxidation area for Mn^{2+} adsorption. At the same time more frequent backwashing cycles could negatively influence development of catalytic layer on the filter media.

A closer look at the inventory data revealed that each of the outliers that showed poor manganese removal at low iron loadings had at least one other parameter that are critical for manganese removal including:

- pH < 7.1 (discussed under *"effect of (filtrate) pH"*) or
- NH_4^+ removal efficiency < 85%, or
- filtration rate > 10.5 m/h and/or EBCT < 11.5 min (discussed under *"effect of filtration rate"* and *"effect of contact time and filter bed depth"*).

2.4.4 Effect of (filtrate) pH

Both PCA and single correlation analysis of the inventory data show that pH has large effect on manganese removal efficiency (Fig. 2.8).

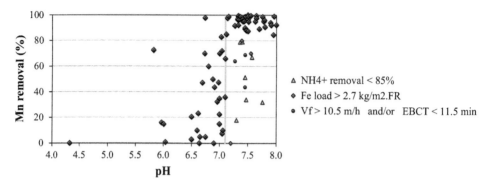

Figure 2.8: *Manganese removal efficiency as a function of filtrate pH.*

The data presented in Fig. 2.8 show that 7.1 is the limiting pH value, below which complete manganese removal was not achieved on any of the plants included in the inventory, with the exception of one specific groundwater treatment plant where complete manganese removal is achieved at a somewhat lower filtrate pH of 6.8. Closer data inspection of this specific plant revealed that the aerated water had a pH > 7.1, and that pH was strongly reduced in the filter bed, probably due to the oxidation of Fe^{2+} (15 mg/L), which resulted in the release of protons. NH_4^+ which can also lower pH, is hardly present in raw water (0.1 – 0.25 mg/L). It is very likely that the pH in the upper part of filter bed was around 7.1, which was sufficient for effective manganese removal, although in this zone iron might compete with manganese for adsorption sites. More research is, however, needed to explain efficient manganese removal archived at pH <7.1 in this plant. Also in literature complete (biological) manganese removal at low pH of 6.5 is reported (Burger *et al.,* 2008). However, it should be mentioned that raw water quality used in the study of Burger *et al.* (2008) differs significantly from raw groundwater quality of plants included in this study (extremely low alkalinity, absence of iron and ammonia, high redox potential). Data shown in Fig. 2.8 strongly indicate that increasing pH, in general, improves manganese removal. Higher pH will strongly enhance adsorption of Mn^{2+} (Buamah, 2009), and will support heterogeneous and autocatalytic manganese adsorption and oxidation. At low pH values, it is known from practice and reported in the literature that physicochemical manganese removal seems to be impossible in traditional aeration-filtration treatment plants. Ramstedt *et al.* (2002) showed that γMnOOH (manganite) starts to dissolve significantly at pH below 6, causing leaching of Mn^{2+} from the adsorbent into the filtrate. Hastings and Emerson (1986) and Klewicki and Morgan (1999) described these dissolution processes as disproportion reactions (Eqs. 2.1, 2.2).

$$Mn_3O_4 \quad + \quad 2H^+ \quad \leftrightarrow \quad 2MnOOH \quad + \quad Mn^{2+} \tag{2.1}$$
$$2MnOOH \quad + \quad 2H^+ \quad \leftrightarrow \quad MnO_2 \quad + \quad Mn^{2+} \quad + \quad 2H_2O \tag{2.2}$$

From Fig. 2.8 it is also clear that manganese removal is very effective at pH values between 7.1 and 8.0. Graveland (1971) and Graveland and Heertjes (1975) reported that there was no substantial oxidation of Mn^{2+} with oxygen in water (homogeneous oxidation) at pH < 8.6 when no catalyst is present. Hence, it is reasonable to expect that the mechanism responsible for manganese removal in treatment plants included in this study is (i) either biological or (ii) a heterogeneous and autocatalytic manganese adsorption and oxidation that takes place on the surface (*e.g.*, MnO_x) of the filter media (Katsoyiannis and Zouboulis, 2004). Fig. 2.8 depicts that manganese removal was less effective on a number of plants included in the inventory where the filtrate pH was higher than 7.1. Close inspection of the data for these plants revealed that at least one of the following parameters limited manganese removal efficiency:

- NH_4^+ removal efficiency $< 85\%$ or
- iron loading per filter run > 2.7 kg Fe / m^2 or
- filtration rate > 10.5 m/h and/or EBCT < 11.5 min (discussed under "*effect of filtration rate*" and "*effect of contact time and filter bed depth*").

2.4.5 Other parameters

According to the single correlation analysis (Table 2.1), the rest of the variables included in this study did not show a significant correlation with the efficiency of manganese removal. However, PCA demonstrates that some parameters (*e.g.*, filtration rate, empty bed contact time and to a lesser extent oxygen and phosphate) may influence manganese removal (Table 2.1 and Fig. 2.4). When plotting a number of parameters such as filtration rate, empty bed contact time (EBCT), and oxygen concentration, against manganese removal efficiency, some trends may indeed be deduced. The effect of these three parameters is discussed in the following paragraphs.

2.4.6 Effect of filtration rate (m/h)

In Fig. 2.9, the manganese removal efficiency obtained on the plants included in the study is plotted as a function of applied filtration rate.

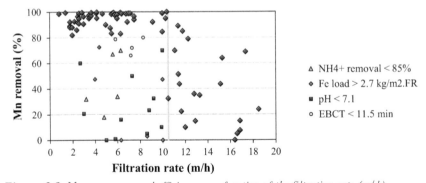

Figure 2.9: *Manganese removal efficiency as a function of the filtration rate (m/h).*

Based on the results from the inventory, complete manganese removal was achieved only in treatment plants that operate at filtration rates of up to 10.5 m/h. Ineffective manganese removal was, however, observed at several treatment plants that operate at filtration rates significantly lower than < 10.5 m/h. These outliers are most likely due to:

- poor NH_4^+ removal efficiency ($< 85\%$) or
- high iron loading (> 2.7 kg Fe / m^2/filter run) or
- low (filtrate) pH (< 7.1) or

- short EBCT (< 11.5 min) (discussed further under "*effect of contact time and filter bed depth*").

Lower filtration rates provide longer contact time in the filter bed, and consequently more time for manganese adsorption and oxidation.

2.4.7 Effect of contact time and filter bed depth

Fig. 2.10 shows manganese removal efficiency as a function of the empty bed contact time (EBCT) for groundwater treatment plants included in the study. It can be seen that only plants that apply an EBCT of 11.5 minutes or more achieved complete manganese removal.

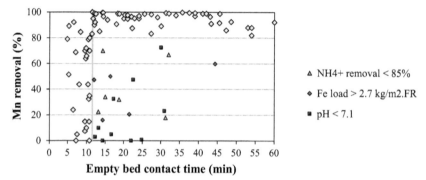

Figure 2.10: *Manganese removal efficiency as a function of the empty bed contact time (EBCT) in minutes.*

Results from the inventory also show that an EBCT longer than 11.5 minutes does not guarantee complete manganese removal. Detailed analysis of plants that apply an EBCT > 11.5 minutes and have ineffective manganese removal showed that again one or more earlier identified water quality parameters and/or process conditions were probably responsible for poor manganese removal.

A parameter directly related to EBCT is bed height of the filter media. No significant statistical correlation between bed height and manganese removal efficiency was found (Table 2.1).

Assuming that a groundwater treatment plant is operating at a maximum acceptable filtration rate of 10.5 m/h, a filter bed of approximately 2.0 m is required to provide a minimal required EBCT of 11.5 minutes that will allow an effective manganese removal. Selected data from the inventory, however, demonstrate that a relatively shallow filter bed height of 0.9 m might be sufficient to obtain complete Mn^{2+} removal, if the other parameters of relevance for manganese removal are optimal (*e.g.*, high pH, low iron loading, low filtration rate, long EBCT).

2.4.8 Effect of oxygen concentration

Only five groundwater treatment plants included in the inventory had dissolved oxygen < 1 mg/L. None of these plants achieved effective manganese removal (Fig. 2.11).

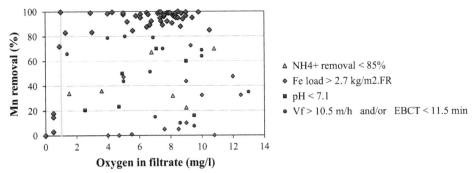

Figure 2.11: *Manganese removal efficiency as a function of dissolved oxygen (mg/L) in filtrate.*

It is known from the literature that manganese removal is negatively affected by the absence of, or low concentration of, *i.e.,* dissolved oxygen (Stumm and Morgan, 1996). Lack of oxygen creates a reducing environment, Mn^{2+} cannot be oxidized, and already formed manganese oxides will start leaching Mn^{2+}. Consequently, a minimal concentration of dissolved oxygen in water is required to prevent manganese oxide reduction, and to support manganese removal by adsorption and oxidation.

Apart from a minimum value of dissolved oxygen concentration required to achieve effective manganese removal, no other distinct relationship between manganese removal efficiency and the oxygen level in filtrate was observed from univariate correlation analysis. Operation of GWTPs with high to very high dissolved oxygen concentrations does not necessarily lead to complete manganese removal. This observation is backed up by the relatively high loading coefficient in the second principal component obtained by PCA (see Table 2.1), indicating that oxygen levels only play an indirect role in manganese removal. More detailed assessment of the data from plants where manganese removal was ineffective with dissolved oxygen in filtrate >1 mg/L revealed that one or more of earlier identified critical water quality or process parameters may explain poor manganese removal. In groundwater, manganese is typically present in low concentrations and exhibits low oxygen demand (0.29 mg O_2 per mg Mn). However, although the oxygen requirement for oxidation of dissolved manganese is very limited, groundwater entering the filters should have a high dissolved oxygen concentration to enable oxidation of dissolved iron (0.14 mg O_2 per mg Fe^{2+}), NH_4^+ (3.6 mg O_2 per mg of NH_4^+), and in some cases methane (4.0 mg O_2 per mg CH_4) and hydrogen sulfide (1.9 mg O_2 per mg H_2S), since these compounds are typically oxidized prior to manganese, if not removed during the aeration (CH_4 and H_2S).

2.4.9 Effect of filtration type (gravity or pressure)

The type of manganese removal filters applied (gravity or pressure filters) was not included in the statistical analysis. Additional assessment of available data revealed that the type of filters applied, could have an important impact on the manganese removal efficiency. In Figure 2.12 (A to D), the effect of some selected water quality and process parameters on manganese removal efficiency is given, taking into account the type of filters applied.

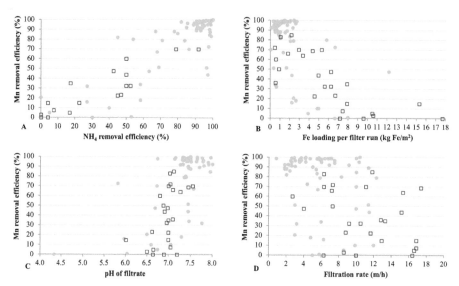

Figure 2.12: *Mn removal efficiency achieved in gravity (●) and pressure filters (□) in conventional aeration-filtration GWTPs as a function of **A:** NH₄⁺ removal efficiency. **B:** Fe loading per filter run (kg Fe/m²). **C:** pH of filtrate. **D:** filtration rate (m/h).*

The results of the inventory show that a large number of GWTPs utilizing gravity filters achieved complete manganese removal, while none of the plants having pressure filters achieved that. It should be noted that pressure filters typically operate at a significantly higher filtration rate (Fig. 2.12D), which results in higher iron loading and shorter contact time (Fig. 2.12B). However, also pressure filters that were operated at a filtration rate < 10.5 m/h, did not achieve complete manganese removal possibly due to a somewhat lower pH (Fig. 2.12C) that could be caused by pressure aeration and (related) lack of CO_2 degassing.

2.5 Conclusions

The manganese removal efficiency of (the first stage of) aeration-rapid sand filtration in GWTPs was assessed. The results from the conducted statistical analysis and assessment of data collected confirmed that manganese removal in the conventional aeration-rapid sand filtration treatment plants is complex and simultaneously influenced by several water quality and process parameters. Close inspection of the collected data from a large number of groundwater treatment plants and conducted statistical analysis (univariate correlations and PCA) shows that manganese removal efficiency is influenced by several parameters simultaneously, including both water quality matrix and several (operational) design parameters. The results of the PCA show that at least four components are necessary to explain at least 75% of the total variance in the data. Furthermore, the PCA reveals that iron loading, NH₄⁺ removal efficiency and pH in filtrate play a major role in manganese removal, while oxygen in the filtrate, and phosphate, manganese and NH₄⁺ concentrations in raw water were found to influence manganese removal to a smaller degree.

Univariate statistics and assessment of available data indicate that very effective manganese removal efficiency in the first aeration-filtration stage can be achieved under the following conditions:

NH$_4$$^+$ removal efficiency	:	> 85%
iron loading per filter run	:	< 2.7 kg Fe/m^2
pH of filtrate	:	> 7.1
filtration rate	:	< 10.5 m/h
empty bed contact time	:	> 11.5 min
oxygen in filtrate	:	≥ 1 mg/L

The results of this study might be a helpful tool for engineers to design a traditional aeration-filtration treatment scheme for proper manganese removal. In addition results emerging from univariate statistics and assessment of available data from full scale plants could be of help in making a decision if manganese removal could be combined with removal of iron and ammonium in a single filtration step.

2.6 Acknowledgements

This research was financially and technically supported by WLN and the Dutch water companies Waterbedrijf Groningen and Waterleiding Maatschappij Drenthe. A large number of Dutch water supply companies Brabant Water, Dunea, Evides, Oasen, Vitens and WML made this study possible by sharing data from their plants. We are also grateful to the Belgian water companies PIDPA and VMW, and the German water companies TAV, SWE, WVO and WAZ Niedergrafschaft, Water supply company of Novi Sad, Serbia and Water Authority of Jordan.

2.7 References

Bartlett, M. S. 1937. "Properties of sufficiency and statistical tests". Proceedings of the Royal Statistical Society, Series A 160, 268–282.

Buamah, R., Petrusevski, B., Schippers, J.C. 2008. Adsorptive removal of manganese (II) from the aqueous phase using iron oxide coated sand. Journal of Water supply: Research and Technology-AQUA 57.1, 1-11.

Buamah, R. 2009. Adsorptive removal of manganese, arsenic and iron from groundwater. PhD thesis, Wageningen University and UNESCO-IHE, The Netherlands.

Buamah, R., Petrusevski, B., de Ridder, D., van de Watering, S. and Schippers, J.C. 2009. Manganese removal in groundwater treatment: practice, problems and probable solutions. Journal of Water Science and Technology: Water Supply 9.1, 89 - 98.

Burger, M.S., Krentz C.A., Nercer S.S., Gagnon G.A. 2008. Manganese removal and occurrence of manganese oxidizing bacteria in full-scale biofilters. Journal of Water Supply: Research and technology-AQUA 57.5, 351-359.

Cerny, C.A., & Kaiser, H.F. 1977. A study of a measure of sampling adequacy for factor-analytic correlation matrices. Multivariate Behavioral Research, 12 (1), 43-47.

Flemming, H. C., Steele, H., Rott, U., Meyer C. 2004. Optimirung der in-situ reaktortechnologie zur dezentralen trinkwassergewinnung und grundwasseraufbereitung durch modelhafte untersuchungen beteiligter biofilme. Report by the Institute for Sanitary Engineering, Water Quality and Solid Waste Management of the University of Stuttgart.

Gouzinis, A., Kosmidis, N., Vayenas, D.V., Lyberatos, G. 1998. Removal of manganese and simultaneous removal of NH_3, Fe and Mn from potable water using trickling filter. Wat. Res. Vol. 32 (8), 2442-2450.

Graveland, A. 1971. Removal of manganese from groundwater, PhD thesis, Technical University Delft, The Netherlands.

Graveland, A., Heertjes, P.M. 1975. Removal of manganese from groundwater by heterogeneous autocatalytic oxidation, Trans. Instn. Chem. Engrs., 53, 154-164.

Hastings, D., Emerson, S. 1986. Oxidation of manganese by marine bacillus: Kinetic and thermodynamic considerations. Geochimica et Cosmochimica Acta 50, 1819-1824.

Huysman, K. 2010. Provinciale en Intercommunale Drinkwatermaatschappij der Provincie Antwerpen (PIDPA), personal communication, Belgium.

Katsoyiannis, I. A., Zouboulis A.I. 2004. Biological treatment of Mn (II) and Fe (II) containing groundwater: kinetic considerations and product characterization. Water research 38, 1922-1932.

Klewicki, J.K., Morgan J.J. 1999. Dissolution of βMnOOH particles by ligands: pyrophosphate, ethylenediaminetetraacetate and citrate. Geochim. Cosmochim. Acta, 63, 3017-3024.

O'Melia, C.R., Crapps, D.K. 1964. Some chemical aspects of rapid sand filtration. Journal of American Water Works Association, October 1964, 1326-1344.

Hu P-Y, Yung-Hsu Hsieh, Jen-Ching Chen, Chen-Yu Chang. 2004a. Adsorption of divalent manganese ion on manganese coated sand, Journal of Water supply: Research and Technology - AQUA 53.3, 151-158.

Hu, Y-H., Hsieh, Y-H., Chen, J-C., Chang, C-Y. 2004b. Characteristics of manganese-coated sand using SEM and EDAX analysis. Journal of Colloid and interface science, 272, 308-313.

Post, J. E. 1999. Manganese oxide minerals: Crystal structures and economic and environmental significance. Proceedings of the National Academy of Sciences USA, 96, 3447-3454.

Ramstedt, M., Schukarev, A., Sjoberg, S. 2002. Characterization of hydrous Manganite (γMnOOH) surfaces – an XPS study. Surf. Interface Anal. 34, 632-636.

Scherer and Wichmann 2000. Treatment of groundwater containing methane – combination of the processing stages desorption and filtration. Acta Hydrochemica et hydrobiologica, 28 (3), 145-154.

Stumm, W. and Morgan, J.J. 1996. Aquatic Chemistry, chemical equilibria and rates, 3rd ed. Wiley, New York.

UNEP 2000. Vital water graphics, http://www.unep.org/vital water

Vandenabeele, J., Van de Woestyne, M., Houwen, F., Germonpré, R., Vandesande, D., Verstreate, W. 1995. Role of Autotrophic Nitrifiers in Biological Manganese Removal from Groundwater containing Manganese and Ammonium. Microbial Ecology 28, 83-98.

Wolthoorn, A., Temminghoff, E.J.M., Weng, L., Riemsdijk, W.H. van 2004. Colloid formation in groundwater: effect of phosphate, manganese, silicate and dissolved organic matter on the dynamic heterogeneous oxidation of ferrous iron. Applied Geochemistry 19, 611-622.

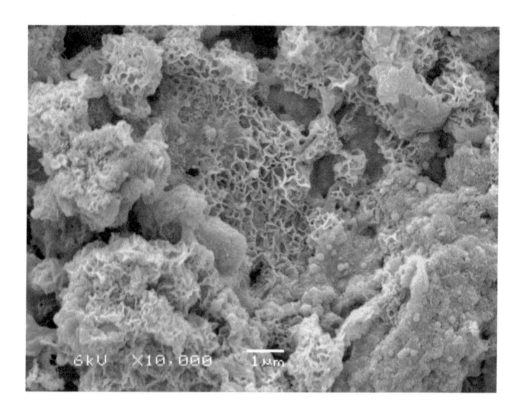

Figure: *Scanning Electron Microscopy (SEM) picture of Birnessite in filter media coating obtained from the pilot research at Grobbendonk - August 2013 (photo: Arie Zwijnenburg, Wetsus)*

3 MANGANESE REMOVAL FROM GROUNDWATER: CHARACTERIZATION OF FILTER MEDIA COATING

Main part of this chapter was published as:
Jantinus H. Bruins, Branislav Petrusevski, Yness M. Slokar, Joop C. Kruithof, Maria D. Kennedy (2015). Manganese removal from groundwater: characterization of filter media coating. Desalination & Water Treatment, 55 (7), 1851 – 1863.

3.1 Abstract

Removal of manganese in conventional aeration-filtration groundwater treatment plants (GWTPs) results in the formation of a manganese oxide coating on filter media. The formation of this coating is an essential prerequisite for efficient manganese removal. Different manganese oxides have varying affinities for autocatalytic adsorption/oxidation of dissolved manganese. The aim of this study was to characterize manganese oxide(s) on filter media from successfully operating manganese removal plants. Characterization of filter media samples from full-scale groundwater treatment plants and identification of manganese species was carried out by X-ray diffraction, scanning electron microscopy coupled with energy dispersive X-radiation (SEM-EDX), Raman spectroscopy and electron paramagnetic resonance (EPR). The results showed that the manganese oxide present in the aged coating was poorly crystalline. Results from the Raman spectroscopy and the detailed EPR analysis show that the manganese oxide in the ripened coating was of a Birnessite type, and of physicochemical origin. The results transpiring from this research suggest that the presence of Birnessite in the coating is essential for effective manganese removal in conventional aeration-filtration treatment plants, since Birnessite has a considerable ability to adsorb and oxidize dissolved manganese.

Keywords: *Groundwater treatment, Manganese removal, MOCS, Filter media characterization, Birnessite*

3.2 Introduction

In European countries the removal of manganese from groundwater is commonly achieved by aeration-rapid sand filtration, eliminating the need for strong oxidants to enhance manganese oxidation. This type of manganese removal is effective and beneficial for both environmental and economic reasons, but requires a long ripening period of virgin filter media. The ripening time can last from several weeks to more than a year, before effective manganese removal is achieved (Cools, 2010; Huysman, 2010; Krull, 2010).

Although extensive research has been carried out on manganese removal by aeration-rapid sand filtration, the mechanisms controlling the ripening period, including the formation of a manganese oxide coating on virgin filter media, are still not well understood. Several researchers have suggested that the use of (pre)coated or ('bio') aged filter media can shorten the ripening period (Buamah *et al.*, 2008; Buamah, 2009; Buamah *et al.*, 2009; Hu *et al.*, 2004a; Islam *et al.*, 2010; Kim *et al.*, 2009; Olanczuk-Neyman *et al.*, 2000; Sahabi *et al.*, 2009; Stembal *et al.*, 2005; Tekerlekopoulou *et al.*, 2005; Tiwari *et al.*, 2007). Aged filter media in manganese removal filters typically consists of Manganese Oxide-Coated Sand (MOCS) and/or Manganese Oxide-Coated Anthracite (MOCA). Autocatalytic properties that are attributed to the coating of these filter media enhance the adsorption of dissolved manganese and its subsequent oxidation. The ripening time of filter media is controlled by the type and amount of manganese oxide(s) present in the coating. As proposed by Stumm and Morgan (1996), oxidation of manganese in homogeneous aqueous solution follows different pathways, as shown in a simplified scheme (Fig. 3.1).

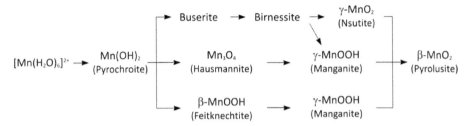

Figure 3.1: *Simplified scheme of Mn²⁺ oxidation pathways according to Stumm and Morgan (1996).*

The propensity of manganese present in filter media coating heavily depends on the valence of manganese in the oxide (Anschutz et al., 2005). A higher valence restricts the possibility of further oxidation. The valences of manganese oxides shown in Fig. 3.1 are given in Table 3.1.

Table 3.1: *Types of manganese oxide and the average valence of manganese in these oxides*

Manganese oxide	Valence of manganese	
Pyrochroite	2	(Stumm and Morgan)
Hausmannite	~ 2.7	(Anschutz *et al.*, 2005)
Manganite / Feitknechtite	3	(Anschutz *et al.*, 2005)
Buserite / Birnessite	3.5 to 3.9	(Cui et al. 2009)]
Nsutite	~ 4	(Anschutz *et al.*,,2005)
Pyrolusite	4	(Anschutz *et al.*, 2005)

In nature, the most stable form of manganese oxide is Pyrolusite (βMnO_2). Because the valence of manganese in this oxide is 4, no further oxidation can take place. Pyrolusite has a large adsorption capacity, but no autocatalytic oxidative properties (Buamah, 2009). Consequently, when the adsorption capacity is exhausted, the removal of manganese stops.

Pyrolusite is formed when (powerful) oxidants such as chlorine (Cl_2), chlorine dioxide (ClO_2), ozone (O_3) or potassium permanganate ($KMnO_4$) are used. These oxidants are commonly applied in countries such as the USA (Carlson et al., 1997; Knocke et al., 1991). Pre-oxidation with powerful oxidants can achieve a very effective manganese removal. However, this type of process is associated with disadvantages such as continuous oxidant dosage to achieve manganese removal. This process is also associated with high costs and environmental risks, and requires accurate oxidant dosage: underdosing causes incomplete manganese oxidation, and consequently a poor manganese removal, while overdosing with permanganate gives the water a pinkish colour. Therefore in Western European countries, such as The Netherlands, manganese removal by conventional groundwater treatment with aeration-filtration is preferred.

Based on thermodynamic considerations, the following pathway for manganese oxidation in water is proposed: (Pyrochroite) – Hausmannite – Manganite – Pyrolusite (Graceland, 1971; Graveland and Heertjes, 1975). However, based on the pɛ / pH diagram shown in Fig. 3.2 (Stumm and Morgan, 1996), chemical formation of Pyrochroite ($Mn(OH)_2$) under common groundwater conditions (pH 6 - 8) is unlikely. Without a catalyst, the pH must be at least 8.6 to achieve Hausmannite (Mn_3O_4) formation, which is subsequently removed by filtration (Graveland, 1971). Formation of Pyrochroite requires an even higher pH, of at least 11.

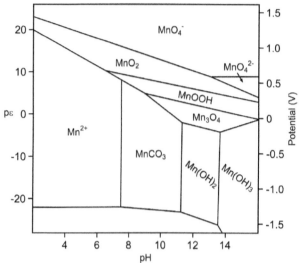

Figure 3.2: *Diagram of electron activity (pε) and redox potential (Eh in V) as a function of the pH showing the stability zones of manganese-containing compounds in aqueous solution (adopted from Stumm and Morgan, 1996).*

The exact pathway by which manganese oxidation occurs under conditions commonly applied in groundwater treatment, is not known. Besides chemical oxidation, 'biology' may play an important role. Identification of the manganese oxide(s) present in filter media coatings from manganese removal filters might give insight into the formation of manganese oxide coatings on virgin filter media. This may elucidate the mechanisms controlling the start-up of the manganese removal process (Hu *et al.,* 2004b). Based on these results, recommendations for process modifications to shorten the ripening period of manganese removal filters may be made.

The aim of this study was to characterize the coating of MOCS and MOCA from (ripened) manganese removal filters in full-scale groundwater treatment plants (GWTPs), by identification of the types of manganese oxide present in the coating. It was hypothesized that identification of the manganese oxide present in ripened media from manganese removal filters may help to better understand the driving force behind efficient manganese removal in conventional aeration-filtration groundwater treatment plants.

3.3 Materials and Methods

MOCS and MOCA characterized in this study were collected from two GWTPs of two water companies as follows.

- MOCS from GWTP De Punt (Groningen, The Netherlands).
  The sample was taken from a depth of 140-150 cm, from a filter bed that was removed after an operating time of approximately 15 years.

- MOCA from GWTP Grobbendonk (Pidpa, Belgium).
  The sample was taken from the top of a filter bed, in operation for more than 12 years.

Both GWTPs utilize conventional groundwater treatment by aeration-filtration and achieve complete manganese removal. Information on feed water quality, process design parameters and operational conditions of full-scale plants De Punt and Grobbendonk are shown in Table 3.2.

Table 3.2: *Feed water quality, process design parameters and operational conditions of full scale GWTPs De Punt and GWTP Grobbendonk*

Parameters feed water / technical information filter	Unit	De Punt	Grobbendonk
Iron	mg/L	4.5 - 6.9	0.03 - 0.14
Manganese	mg/L	0.18 - 0.25	0.12 - 0.18
Ammonium	mg/L	0.29 - 0.78	< 0.05 - 0.23
pH	[-]	7.3 -7.5	7.5 - 7.6
Oxygen	mg/L	8 - 10	> 10
Redox potential	mV	-50 to +50	+ 200 to +300
Type of aeration	-	spray	cascade
Position of filter	-	pre-filter	post-filter
Type of filtration	-	down flow	down flow
Type of filter media	-	quartz sand	anthracite / quartz sand
Grain size fraction virgin media	mm	1.8 - 2.4	0.8-1.8 / 0.4-0.8
Filter area	m²	12.5	37.5
Filter bed height	m	2	1.1 (0.6+0.5)
Flow per filter	m³/h	60	190
Filtration rate	m³/m².h	4.8	5.0
Empty bed contact time	min	25	13.2
Backwash (BW) criterion	-	head loss	head loss
Backwash frequency	n/week	2	0.5
Filter bed expansion during BW	-	no	yes (anthracite)
Filtered volume between BW	m³ per filter run	5,000 – 7,000	10,000
Iron loading per filter run (FR)	kg Fe/m².FR	2.5	< 0.1

To characterize the filter media coatings and to identify the manganese oxide(s), the following supplementary techniques were used:

- Raman spectroscopy;
- X-ray Diffraction (XRD);
- Scanning Electron Microscopy coupled with Energy Dispersive X-radiation analysis (SEM-EDX) and
- Electron Paramagnetic Resonance (EPR).

As these techniques are supplementary, for complete identification and determination of the origin of produced MnO_x, application of all these methods is required. Use of one single method is insufficient to characterize and identify the manganese oxides.

For additional information with respect to Raman spectroscopy (selection spectroscopy settings - Section 3.1) and SEM-EDX (measuring trace (counter) elements) Section 3.3), additional MOCS samples from post filters of two other plants were used, as follows.

- MOCS from GWTP Onnen (Groningen, The Netherlands).
  The sample is taken from the top of the filter, in operation for more than 40 years.

- MOCS from GWTP Wierden (Vitens, The Netherlands).
  The sample is taken from the top of the filter, in operation for more than 18 years.

3.3.1 Raman spectroscopy

With Raman spectroscopy it is possible to distinguish different (general) types of manganese oxides. For the Raman spectroscopy analysis a Horiba Yobin Yvon Labram instrument was used with the following settings (Table 3.3).

Table 3.3: *Parameter settings for Raman spectroscopy analysis.*

Parameter	Setting
Exposure	≥ 30 sec
Current	0.05 mA
Confocal hole	1,000 μm
Slit	100 μm
Laser wave length	532.13 nm
Grating	600
Objective	x50
Density filter	D3
Detector	synapse CCD
Detector size	1,024 pixels

Exposure of the samples to high power laser radiation (5 mA) during Raman analysis for a long time (> 120 seconds) may change the structure of the manganese oxides. However, under the applied conditions (Table 3.3), no structural damage was detected during the tests. Detailed information on the effect of high power radiation on structural changes of the manganese oxides is given in Section 3.4.1. Before analysis, the samples (either the integral grain or powder coating) were dried at room temperature to avoid excessive fluorescence caused by the presence of water in the samples.

3.3.2 XRD

XRD, can be used to determine whether manganese oxide is crystalline or amorphous. Furthermore, this technique can provide additional information to clarify the sharpness of the peaks found with Raman spectroscopy. Most XRD measurements were carried out with a Bruker-D8 Advance diffractometer in Bragg-Brentano focusing geometry. The instrument was equipped with a Vantec PSD detector. Measurements were carried out at room temperature with the use of monochromatic CoKα1 radiation ($\lambda =$ 0.179026 nm). The 2θ scan was made in the range 10-110°2θ, using a step size of 0.038°2θ. To check some of the analyses and to enhance their performance, some XRD measurements were also carried out with a Bruker-AXS D5005 diffractometer in Bragg-Brentano focusing geometry, equipped with a graphite monochromator in the diffracted beam. Before analysis, the samples were dried at room temperature and then pulverized to a size of 10-50 μm. The sample powder (for MOCS only the coating and for MOCA the coating including the anthracite grain) was put in an aluminum sample holder. The 2θ scan was made in the range 10-110°2θ, using a step size of 0.025°2θ and a counting time of 2 s per step. The radiation was CuKα1 (0.15406 nm). Results obtained by both the XRD instruments were evaluated with the internal Bruker EVA software.

3.3.3 SEM-EDX

SEM pictures show the structure of the coating and the manganese oxide inside. In addition EDX provides information about the trace (counter) elements present in these structures. Therefore, differences in peak ratio found with Raman spectroscopy can be clarified. The SEM pictures were made with a JEOL-6480LV. Samples (either the integral grain or powder coating) were placed on 15 mm diameter mounts with a double-sided carbon adhesive tab. The SEM was operated both under High Vacuum (HV) and Low Vacuum (LV) conditions.

Before observation under HV conditions, the samples were coated in a JEOL JFC-1200 fine coater with a thin (10 nm) Au layer. The operation under HV conditions was done at 6 kV at 10 mm Working Distance (WD) and Spot Size (SS) 20. Samples were also investigated without Au coating under LV conditions at 10 kV and SS 60. Composition of the adsorbed coating layers was determined with an X-ray microanalysis (EDX) System type Noran System SIX from the Thermo Electron Corporation. Analyse were done at 10 kV and SS 70 for both the Au-coated samples at HV conditions and non-coated samples at LV conditions. In the latter, no Au peaks were observed in the EDX spectrum.

3.3.4 EPR

Once the MnO_x is characterized as 'a type of Birnessite', EPR can be employed to identify the exact type of Birnessite and additionally also to determine the origin of the Birnessite. For EPR the whole grain was always used.

EPR analyses were carried out with a Bruker instrument operating at 9.46 GHz (W-waveband). The analyses were carried out at two different temperatures, *i.e.*, at 77 K (cooled with nitrogen) and 298 K (room temperature).

3.4 Results and Discussion

3.4.1 Selection Raman spectroscopy settings

As mentioned in Section 3.3.1, the structure of almost all manganese (oxide) compounds (except Hausmannite) is subjected to Raman spectroscopic changes when exposed to a high current (Julien et al., 2003, 2004). Laser irradiation in excess of 5 mA for longer than 120 s causes structural changes of the exposed compounds (polymorphism). To prevent these changes the current of the laser in this study was limited to 0.05 mA. The differences in spectra obtained with high (5 mA, >120 s) and low (0.05 mA) radiation intensities are shown in Fig. 3.3.

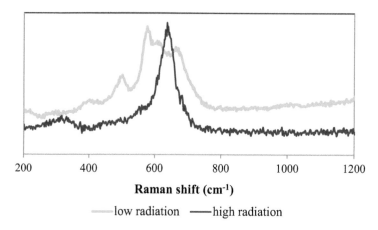

low radiation ——high radiation

Figure 3.3: *Raman spectra at 532 nm of MOCS (Onnen), exposed to low (light blue line) and high laser radiation (red line).*

From the two spectra in Fig. 3.3 it can be seen that high intensity laser irradiation changes the position of the peaks, as well as the pattern of the spectrum. According to (Julien *et al.,* 2003, 2004), high current irradiation transforms all MnO_x samples (and therefore also Birnessite) into Mn_3O_4 (Hausmannite). Hausmannite is very stable when exposed to a Raman laser, so its structure is not changed by the high level of irradiation. Furthermore, due to its crystalline lattice structure it displays a sharp peak in the Raman spectrum. To ascertain this transformation, Raman profiles of Birnessite from this study and an MnO_2 reference sample (Alfa Aesar) were exposed to high current radiation (5 mA, >120 s), and compared to a Hausmannite spectrum from the RRUFF database (Downs, 2006) (Fig. 3.4).

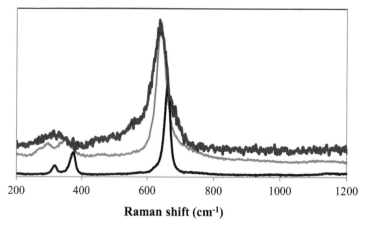

Figure 3.4: *Raman spectra (at 532 nm) of Birnessite from a sample of this study (red line) and MnO_2 reference (green line) after exposure to 5 mA for >120 s, compared to a Hausmannite spectrum (purple line) from the RRUFF database (Downs, 2006).*

From Fig. 3.4 it can be seen that the spectra of Birnessite in a sample from this study as well as the MnO$_2$ reference, exposed to high laser radiation, show a close resemblance to the spectrum of Hausmannite. This indicates that both manganese oxides (Birnessite and MnO$_2$) underwent structural changes when exposed to high laser power, as stated by (Julien *et al.*, 2004).

3.4.2 Characterization of MOCA and MOCS by Raman spectroscopy

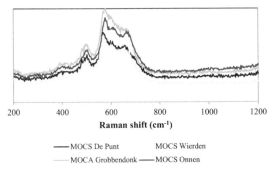

Figure 3.5: *Raman spectra (at 532 nm) of the four MOCS / MOCA samples.*

The Raman spectra of manganese-coated samples from four GWTPs are presented in Fig. 3.5. From Fig. 3.5, it can be seen that the spectral patterns of all samples were similar, indicating a similar type of manganese oxide. The three most pronounced absorbance peaks were found at Raman shifts of 495-505 cm^{-1}, 570-575 cm^{-1}, and 635-655 cm^{-1}, typical for the Birnessite group of manganese oxides (Julien, *et al.*, 2003). All spectra exhibited a less pronounced peak at a Raman shift at 400-420 cm^{-1}. The undulating peaks were typical for oxides with a poorly crystalline structure, such as Birnessite (Post, 1999).

To confirm the presence of Birnessite, the spectra of the MOCS/MOCA samples were compared with the spectra of synthetically produced Birnessite (Ma *et al.*, 2007), shown in Fig. 3.6A. Comparison of the spectra produced in this study with previously reported results strongly suggested that the manganese oxide in the coating of the MOCS and MOCA samples was of a Birnessite type.

The spectra of MOCS and MOCA samples were also compared with a Raman profile of a reference sample of naturally formed Birnessite (Fig. 3.6B), obtained from the Aufgeklärt Glück mine in Hasserode, the Harz Mountains, Sachsen-Anhalt, Germany (Witzke, 2012).

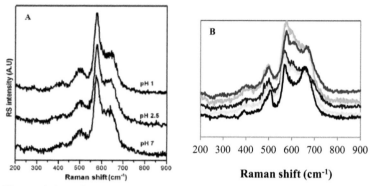

Figure 3.6: *A: Raman spectra of synthetically produced Birnessite at 514 nm (Ma et al., 2007). B: Raman spectra of samples from this study (green, red, blue and yellow lines) compared to the spectrum of the Birnessite reference (black line).*

Fig. 3.6B shows that the main peaks from the Birnessite reference as well as the MOCA and MOCS samples were on the same Raman shift. Only the ratio and height of the peaks varied between the different samples and the reference, probably due to a different concentration of counter ions in the materials (Julien *et al.*, 2003, 2004), as shown by SEM-EDX (Section 3.4.4, Table 3.4).

The Raman spectra strongly suggested that the manganese oxides in the MOCA and MOCS samples were of a Birnessite type.

In order to exclude the presence of other manganese oxides in the coatings, the Raman spectra of samples from this study were compared with the reference spectra of four other manganese oxides Pyrochroite (A), Hausmannite (B), Manganite (C) and Pyrolusite (D) (Fig. 3.7). Spectra A-C were taken from the RRUFF database (Downs, 2006) and spectrum D from the measured reference sample (Alfa Aesar).

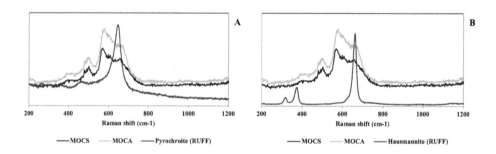

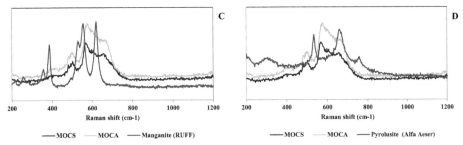

Figure 3.7: *Raman spectra at low power and 532 nm of MOCS De Punt and MOCA Grobbendonk, compared to (A) Pyrochroite, (B) Hausmannite, (C) Manganite and (D) Pyrolusite.*

The four manganese oxides displayed one peak in the same region as the MOCS/MOCA samples between 630 and 650 cm^{-1}. This wavelength is characteristic for all manganese oxides and not indicative of a particular one. Therefore, it was concluded that apart from Birnessite no other manganese oxides were present in the coating of the filter media.

3.4.3 XRD

Fig. 3.8 shows an XRD spectrum of MOCS coating from GWTP De Punt and a MOCA coating from GWTP Grobbendonk. The analyzed MOCA sample contained some crushed anthracite (carbon) since it was impossible to separate the coating completely from the anthracite.

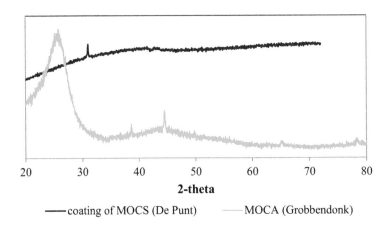

Figure 3.8: *XRD spectra (conducted with a Bruker D8), of pulverized MOCS coating and MOCA (the latter including crushed anthracite core).*

From Fig. 3.8 it can be seen, as already shown by the Raman spectroscopy, that the MOCS coating was poorly crystalline. The only peak (31^0, 2θ) belonged to silica originating from the filter media. Additional (sharp) peaks were not present in the spectra from the MOCS sample, so no crystalline MnO_x was present on the MOCS. Also no crystalline MnO_x was found on the MOCA. The few small peaks and the broad peak in the MOCA sample (25^0, 2θ) originate from carbon or graphite (C) from crushed anthracite that could not be removed completely from the coating (as mentioned above).

For comparison, the XRD spectra of three Alfa Aesar reference manganese oxides *i.e.*, MnO_2, Mn_2O_3, and MnO with crystalline structure are presented in Fig. 3.9.

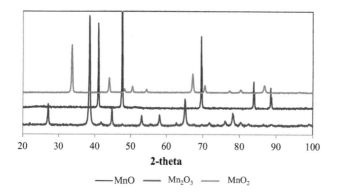

2-theta

——MnO —— Mn_2O_3 —— MnO_2

Figure 3.9: *XRD spectra of reference manganese oxides: MnO, Mn₂O₃ and MnO₂ (Alfa Aesar).*

Contrary to the MOCA and MOCS coating, the XRD spectra of the three reference manganese oxides showed sharp peaks and the 2θ positions of the peaks were confirmed by the reference XRD spectra of oxides from the internal Bruker EVA database. If the manganese oxide in the samples from this study had been crystalline, sharp peaks should have been observed similar to the spectra of the three reference manganese oxides. The absence of sharp peaks confirmed the amorphous character of the sample coating. Thus the XRD results also confirmed that the manganese oxide(s) in the MOCS and MOCA samples were not crystalline manganese oxides (*i.e.*, MnO_2, Mn_2O_3 or MnO).

3.4.4 SEM-EDX

Figure 3.10: *SEM image of virgin quartz media (sand); 500× magnification.*

In Fig. 3.10 a SEM image of virgin quartz (sand) filter media is shown. Fig. 3.10 shows that the surface of virgin sand was not completely smooth, indicating a high porosity and specific surface area, and therefore probably contained many sites for attachment of bacteria and/or manganese, which could shorten the ripening time.

In Fig. 3.11 SEM images of MOCS and MOCA samples are shown.

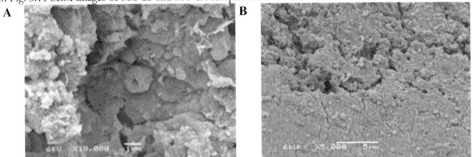

Figure 3.11: *SEM images of filter media coating. A - MOCS De Punt; 10,000x. B - MOCA Grobbendonk; 5,000x.*

Images A and B in Fig. 3.11 confirm the Raman spectroscopy results and the XRD analysis that the manganese coating was poorly crystalline.

Results of the SEM-EDX analyses show that apart from manganese and iron, the coating of the filter media contained (trace) elements such as aluminum, calcium, magnesium, potassium, sodium and/or silica (Table 3.4).

Table 3.4: *SEM-EDX analysis of counter (trace) elements in weight (%).*

Element	MOCA Grobbendonk	MOCS De Punt	MOCS Onnen	MOCS Wierden
Aluminum (Al)	0.5	< 0.1	0.6	< 0.1
Calcium (Ca)	6.8	2.3	7.2	7.7
Magnesium (Mg)	0.3	< 0.1	0.4	0.3
Potassium (K)	0.2	< 0.1	< 0.1	< 0.1
Silica (Si)	1.0	5.9	2.6	0.9
Sodium (Na)	< 0.1	< 0.1	0.2	< 0.1

The counter (trace) elements in Table 3.4 are of importance with respect to the ratio and height of the Raman peaks, as discussed in Section 3.4.2 and shown in Fig. 6B. Counter ions present in filter media coatings may also impact the adsorptive properties of the coating.

3.4.5 EPR

In Fig. 3.12 the EPR spectra of MOCS De Punt measured at two temperatures (298 K - room temperature and 77 K) are shown. Due to the analytical interference of carbon (originating from anthracite) it was not possible to obtain a complete EPR spectrum for MOCA Grobbendonk.

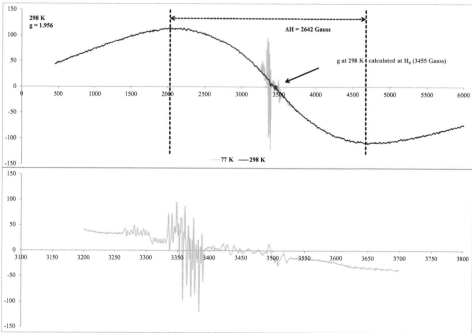

Figure 3.12: *EPR spectra of MOCS De Punt. Top half: spectra measured at 298 K (blue line) and 77 K (green line). Bottom half: expanded spectrum measured at 77 K.*

The detailed EPR pattern (Fig. 3.12, the bottom half) shows the '6 line hyperfine structure of manganese'. This pattern is typical for oxides containing manganese ions with a different valence, including Mn^{2+} (Browne, 2011). Birnessite has a valence between 3.5 and 3.9 (Cui *et al.*, 2009; Post, 1999; Kim *et al.*, 2011), due to the presence of Mn^{3+} and Mn^{4+} in the lattice. This pattern is an additional indication that the predominant manganese component of the MOCS coating is a manganese oxide, containing manganese ions with different oxidation numbers, such as Birnessite.

Generally, EPR spectra are characterized by two parameters: (i) the line width: ΔH in gauss, and (ii) the spectroscopic splitting factor: g (g-factor). EPR was used to investigate the origin of different types of Birnessite (Kim *et al.*, 2011). These values of ΔH measured at 77 K and 298 K and the g-factor calculated at 298 K are characteristic of Birnessite.

ΔH is measured as the distance between the highest and lowest point of the signal wave, and the g factor is calculated from the spectrometer microwave frequency (y) and the magnetic field (H_0) applied during the analysis when the signal is zero (Eq. 1):

$$g = \frac{h}{\mu_B} \times \frac{y}{H_0}$$

(1)

where:

h	=	Planck constant (6.62×10^{-34} Js);
μ_B	=	Bohr magneton (9.27×10^{-28} J/G);
y	=	spectrometer microwave frequency (GHz);
H_0	=	applied magnetic field where signal is zero (gauss).

46

ΔH values measured at 298 K and at 77 K were 2,642 gauss and 3,950 gauss, respectively (Fig. 3.12 - top half). Taking into account the applied magnetic field (H_0) of 3,455 gauss at 298 K, the calculated g factor was 1.956. Comparing the EPR results from this study with the data reported by (Kim *et al.*, 2011), a few conclusions can be drawn. Firstly, the EPR analysis confirmed the results obtained by Raman spectroscopy that the manganese oxide on MOCS is Birnessite. Next, comparing the two ΔH values and the g factor with the results from Kim *et al.* (2011) shows that the Birnessite in the coating of MOCS De Punt was of physicochemical origin. However, this does not mean that the growth of manganese oxide on the virgin filter media (ripening process) starts exclusively chemically. The MOCS sample from GWTP De Punt was taken after more than 15 years of operation, and consequently the EPR analysis only confirmed the physicochemical nature of Birnessite after this elapsed time.

3.4.6 The importance of Birnessite formation on MOCA/MOCS for manganese removal

The mineral Birnessite was first found in 1956 in a small hamlet called Birness, Scotland, UK (Jones and Milne, 1956). At that time it was a new mineral to which the name Birnessite, after the locality, was given. Its molecular structure was given as $(Na_{0.7}Ca_{0.3})Mn_7O_{14} \cdot 2.8H_2O$. In time, different formulas were proposed for Birnessite, indicating there is no single structure, but different compositions exist ('minerals of the Birnessite type'). From this study it can be observed that Birnessite was the manganese compound formed during filter ripening, and predominantly present in all manganese oxide coatings extracted from the four GWTPs that showed complete manganese removal. Consequently, Birnessite appears to be of particular importance for manganese removal in conventional aeration-filtration groundwater treatment plants. The presence of Birnessite may provide an explanation for the very effective manganese removal in these filters in practice. Birnessite is extremely suitable to remove manganese, because of its structure as described by Post (Post, 1999): 'the Birnessite group of minerals has layered structures, which may readily undergo oxidation-reduction and cation-exchange reactions and play a major role in controlling groundwater chemistry'. The high cation exchange and adsorption capacity of Birnessite is also described by several other researchers (Kim *et al.*, 2009; Golden *et al.*, 1986; Han *et al.*, 2006; Kim *et al.*, 2008; Lee *et al.*, 2009; Murray *et al.*, 1976; Pretorius and Linder, 2001; White, 1997). Thus, Birnessite is very suitable to adsorb Mn^{2+}.

The Birnessite group of minerals (including Buserite) is also important, because of its high reactivity (Post, 1999; Post and Veblen, 1999). In Buserite and Birnessite the average valence number of manganese ranges from +3.5 to +3.9 (Cui *et al.*, 2009). Although restricted compared to other manganese oxides, further oxidation of Buserite and Birnessite is still possible. The combination of properties makes Birnessite a highly reactive manganese oxide, with very good adsorptive properties for dissolved manganese, and its subsequent autocatalytic oxidation. The knowledge that Birnessite is the manganese oxide responsible for efficient manganese removal can help substantially accelerate the ripening process of virgin filter media, by creating conditions favouring the formation of Birnessite. Knowing that the oxidation from Mn^{2+} into Pyrolusite (Stumm and Morgan, 1996), via Birnessite (Pyrochroite $\rightarrow$ Buserite / Birnessite $\rightarrow$ Nsutite $\rightarrow$ Pyrolusite, Fig. 3.2), is only possible under (very) alkaline conditions (Feng *et al.*, 2005), it is not likely that a fast filter ripening of virgin filter media starts in a chemical way, without oxidant dosage. Therefore the formation of the manganese coating may be initiated by bacterial activity. Several researchers have suggested that manganese removal is influenced by bacteria (*e.g.*, *Pseudomons* sp., *Leptothrix* sp.), which are able to oxidize Mn^{2+} (Kim *et al.*, 2011; Barger *et al.*, 2005; Burger *et al.*, 2008a, 2008b; Geszvain, 2011; Katsoyiannis and Zouboulis, 2004; Tebo *et al.*, 2004; Villalobos *et al.*, 2003; Vandenabeele *et al.*, 1992, 1995; Vandenabeele, 1993, Tekerlekopoulou *et al.*, 2008). Therefore the ripening of the filter media probably starts with the biological formation of Birnessite. In time very fast physicochemical auto-catalytic adsorption/oxidation reactions may become more important and result in production of Birnessite of physicochemical origin, whose presence is shown in this study. However, further research is required to support this hypothesis.

3.5 Conclusions

The Raman spectroscopy, XRD and SEM analyses carried out in this study showed that the manganese oxide in the coating of the manganese removing filter media is poorly crystalline. Raman spectroscopy and EPR analysis further clarified that the predominant manganese oxide, responsible for effective removal of dissolved manganese, is of a Birnessite type. Calculation of ΔH and the g factor from EPR analysis and comparison of these parameters with results from literature identified Birnessite as being of physicochemical origin, but the sampling after a ripening period of about 15 years does not exclude the possibility of Birnessite formation starting via a biological pathway. Despite the generally accepted theory that the manganese oxidation pathway is via Hausmannite and Manganite, the results transpiring from this research imply that in water treatment practice oxidation of manganese on the surface of manganese removal filter media is more likely to form a Birnessite type of manganese oxide. Birnessite has very good properties for adsorption and autocatalytic oxidation of dissolved manganese. Identification of Birnessite as the predominant manganese oxide in filter media that effectively remove manganese could possibly enable shortening ripening time in conventional aeration-filtration groundwater treatment plants by creating conditions that favour the formation of this compound.

3.6 Acknowledgements

This research is financially and technically supported by WLN and by the Dutch water companies Waterbedrijf Groningen (WBG) and Waterleiding Maatschappij Drenthe (WMD). The authors of this paper would like to thank Mr. Arie Zwijnenburg and Mr. Ton van der Zande from Wetsus, Centre of Excellence for Sustainable Water Technology, Mr. Wesley Browne from the University of Groningen and Mr. Ruud Hendrikx of the Department of Materials Science and Engineering from the Technical University of Delft for their help with the analyses of samples and interpretation of the results. The authors would also like to thank the Belgian water company Pidpa and the Dutch water company Vitens for providing filter media for this study and their willingness to disclose their groundwater treatment plant data. Finally we would like to thank Dr. Thomas Witzke (mineralogist) for donating a naturally formed Birnessite reference sample.

3.7 References

Anschutz, P., Dedieu, K., Desmzes, F. and Chaillou, G. 2005. Speciation, oxidation state, and reactivity of particulate manganese in marine sediments. Chem. Geol., 218, 265-279.

Barger, J.R, B.M. Tebo, B.M., Bergmann, U., Webb, S.M., Glatzel, P., Chiu, V.Q., and Villalobos, M. 2005. Biotic and abiotic products of Mn (II) oxidation by spores of the marine *Bacillus* sp. strain SG-1. Am. Mineral., 90, 143-154.

Browne, W.R., 2011. Assistant professor, Stratingh Institute for Chemistry, University of Groningen, personal communication, The Netherlands.

Buamah, R., Petrusevski B. and Schippers, J.C. 2008. Adsorptive removal of manganese (II) from the aqueous phase using iron oxide coated sand. J. Water Supply Res. T., 57.1, 1-11.

Buamah, R. 2009. Adsorptive removal of manganese, arsenic and iron from groundwater, PhD thesis, UNESCO-IHE Delft / Wageningen University, The Netherlands.

Buamah, R., Petrusevski, B., de Ridder, D., van de Wetering, S. and Schippers J.C. 2009. Manganese removal in groundwater treatment: practice, problems and probable solutions, Water Sci. Technol., 9.1, 89-98.

Burger, M.S., Mercer, S.S., Shupe G.D. and Gagnon G.A. 2008a. Manganese removal during bench-scale biofiltration. Water Res., 42, 4733-4742.

Burger, M.S., Krentz, C.A., Mercer S.S. and Gagnon G.A. 2008b. Manganese removal and occurrence of manganese oxidizing bacteria in full-scale biofilter. J. Water Supply Res. T., 57.5, 351-359.

Carlson, K.H., Knocke W.R. and Gertig K.R. 1997. Optimizing treatment through Fe and Mn fractionation, J. Am. Water Works Assoc., 89.4, 162-171.

Cools B. 2010. Vlaamse Maatschappij voor Watervoorziening (VMW), personal communication, Belgium.

Cui, H., Qiu, G., Feng, X., Tan, W. and Liu F. 2009. Birnessite with different average manganese oxidations states synthesized, characterized, and transformed to Todorokite at atmospheric pressure, Clay. Clay Miner., 57.6, 715-724.

Downs, R.T. 2006. The RRUFF Project: An integrated study of the chemistry, crystallography, Raman and infrared spectroscopy of minerals, program and abstracts of the 19th general meeting of the international mineralogical association in Kobe, Japan.

Feng, X., Tan, W., Liu, F., Huang Q. and Liu, X. 2005. Pathways of Birnessite formation in alkali medium. Sci. China Ser. D., 48.9, 1438-1451.

Geszvain K., Yamaguchi, Ai., Maybee, J. and Tebo, B.M. 2011. Mn (II) oxidation in *Pseudomonas putida* GB-1 is influenced by flagella synthesis and surface substrate, Arch. Microbiol., 193, 605-614.

Golden, D.C., Dixon, J.B. and Chen, C.C. 1986. Ion exchange, thermal transformations, and oxidizing properties of Birnessite. Clay. Clay Miner., 34.5, 511-520.

Graveland, A. 1971. Removal of manganese from groundwater, PhD thesis, Technical University Delft, The Netherlands.

Graveland, A. and Heertjes P.M. 1975. Removal of manganese from groundwater by heterogeneous autocatalytic oxidation. Trans. Inst. Chem. Eng., 53, 154-164.

Han, R., Zou, W., Zhang, Z., Shi, J. and Yang, J. 2006. Removal of copper (II) and lead (II) from aqueous solution by manganese oxide coated sand. I. Characterization and kinetic study. J. Hazard. Mater, 137, 384-395.

Hu, P.Y., Hsieh, Y.H., Chen J.C. and Chang, C.Y. 2004a. Adsorption of divalent manganese ion on manganese-coated sand, J. Water Supply Res. T., 53.3, 151-158.

Hu, P.Y., Hsieh, Y.H., Chen J.C. and Chang, C.Y. 2004b. Characteristics of manganese-coated sand using SEM and EDAX analysis. J. Colloid Interf. Sci., 272, 308-313.

Huysman, K. 2010. Provinciale en Intercommunale Drinkwatermaatschappij der Provincie Antwerpen (PIDPA), personal communication, Belgium.

Islam, A.A., Goodwill, J.E., Bouchard, R., Tobiasen J.E. and Knocke W.R. 2010. Characterization of filter media $MnO_2(s)$ surfaces and Mn removal capability. J. Am. Water Works Assoc., 102.9, 71-83.

Julien, C., Massot, M., Baddour-Hadjean, R., Franger, S., Bach, S. and Pereira-Ramos, J.P. 2003. Raman spectra of Birnessite manganese dioxides. Solid State Ionics, 159, 345-356.

Julien, C., Massot, M. and Poinsignon, C. 2004. Lattice vibrations of manganese oxides. Part I. Periodic structures. Spectrochim. Acta A, 60, 689-700.

Jones, L.H.P. and Milne, A.A. 1956. Birnessite, a new manganese oxide mineral from Aberdeenshire, Scotland, Mineral. Mag. Journal Min. Soc., XXXI.235, 283-288

Katsoyiannis I.A. and Zouboulis, A.I. 2004. Biological treatment of Mn (II) and Fe (II) containing groundwater: Kinetic considerations and product characterization. Water Res., 38, 1922-1932.

Kim, S.S., Bargar, J.R., Nealson, K.H., Flood, B.E., Kirschvink, J.L., Raub, T.D., Tebo B.M. and Villalobos M. 2011. Searching for bio signatures using electron paramagnetic resonance (EPR) analysis of manganese oxides. Astrobiology, 11.8, 775-786.

Kim, I. and Jung, S. 2008. Soluble manganese removal by porous media filtration. Environ. Technol., 22, 1265-1273.

Kim, W.G., Kim, S.J., Lee, S.M. and Tiwari, D. 2009. Removal characteristics of manganese-coated solid samples for Mn (II). Desal. Wat. Treat., 4, 218-223.

Knocke, W.R., van Benschoten J.E., Kearney, M.J., Soborski A.W. and Reckhow, D.A. 1991. Kinetics of manganese and iron oxidation by Potassium Permanganate and chlorine dioxide. J. Am. Water Works Assoc., 83.6, 80-87.

Krull, J. 2010. Stadtwerke Emden (SWE), personal communication, Germany.

Lee, S.M., Tiwari, D., Choi, K.M., Kim, W.G., Yang, J.K. and Lee, H.D. 2009. Removal of Mn (II) from aqueous solutions using manganese coated sand samples, J. Chem. Eng. Data, 54, 1823-1828.

Ma, S.B., Ahn, K.Y., Lee, E.S., Oh, K.W. and Kim, K.B. 2007. Synthesis and characterization of manganese dioxide spontaneously coated on carbon nanotubes. Carbon, 45, 375-382.

Murray, J. 1976. The interaction of metal ions at the manganese dioxide-solution interface. Geochim. Cosmochim. Ac., 39, 606-619.

Olanczuk-Neyman, K. and R. Bray 2000. The role of physico-chemical and biological processes in manganese and ammonia nitrogen removal from groundwater. Pol. J. Environ. Stud., 9.2, 91-96.

Post, J.E. 1999. Manganese oxide minerals: Crystal structures and economic and environmental significance, P. Natl. Acad. Sci. USA, 96, 3447-3454.

Post, P.E. and Veblen, D.R. 1999. Crystal structure determinations of synthetic sodium, magnesium, and potassium Birnessite using TEM and the Rietveld method.Am. Mineral., Volume 7.5, 477-489.

Pretorius, P.J. and Linder P.W. 2001. The adsorption characteristics of δ-Manganese dioxide: A collection of diffuse double layer constants for the adsorption of H^+, Cu^{2+}, Ni^{2+}, Zn^{2+}, Cd^{2+} and Pb^{2+}. Appl. Geochem., 16, 1067-1082.

Sahabi, D.M., Takeda, M., Suzuki, I. and Koizumi, J.I. 2009. Removal of Mn^{2+} from water by 'aged' biofilter media: The role of catalytic oxides layers. J. Bio. Sci. Bioeng., 107.2, 151-157.

Stembal, T., Markic, M., Ribicic, N., Briski, F. and Sipos, L. 2005. Removal of ammonia, iron and manganese from ground waters of Northern Croatia – pilot plant studies, Process Biochem., 40, 327 -335.

Stumm, W. and Morgan, J.J. 1996. Aquatic chemistry, chemical equilibria and rates, 3rd ed. Wiley, New York, 464-467.

Tebo, B.M., Marger, J.R., Clement, B.G., Dick, G.J., Murray, K.J., Parker, D., Verity, R. and Webb, S.M. 2004. Biogenic Manganese oxides: Properties and mechanisms of formation, Annu. Rev. Earth Pl. Sci., 32, 287-328.

Tekerlekopoulou, A.G., Vasiliadou I.A. and Vayenas, D.V. 2008. Biological manganese removal from potable water using trickling filters. Biochem. Eng. J., 38, 292-301.

Tekerlekopoulou A.G. and Vayenas, D.V. 2008. Simultaneous biological removal of ammonia, iron and manganese from potable water using a trickling filter. Biochem. Eng. J., 39, 215-220.

Tiwari, D., Yu, M.R., Kim, M.N., Lee, S.M., Kwon, O.H., Choi, K.M., Lim G.J.and Yang, J.K. 2007. Potential application of manganese coated sand in the removal of Mn (II) from aqueous solutions. Water Sci. Technol., 56:7, 153-160.

Vandenabeele, J., de Beer, D., Germonpré R. and Verstreate, W. 1992. Manganese oxidation by microbial consortia from sand filters, Microb. Ecol., 24, 91-108.

Vandenabeele, J., van de Woestyne, M., Houwen, F., Germonpré, R., Vandesande D. and Verstreate, W. 1995. Role of autotrophic nitrifiers in biological manganese removal from groundwater containing manganese and ammonium. Microb. Ecol., 28, 83-98.

Vandenabeele, J. 1993. Manganese removal by microbial consortia from rapid sand filters treating water containing Mn^{2+} and NH_4^+, PhD thesis, Gent, Belgium.

Villalobos, M., Toner, B., Barger J. and Sposito, G. 2003. Characterization of the manganese oxide produced by *Pseudomonas putida* strain MnB1. Geochim. Cosmochim. Ac., 67(14), 2649-2662.

White D.A. and Asfar-Siddique, A. 1997. Removal of Manganese and iron from drinking water using hydrous Manganese Dioxide. Solvent Extr. Ion Exc., 15(6), 1133-1145.

Witzke, T. 2012. mineralogist PANanalytical, personal communication, Germany.

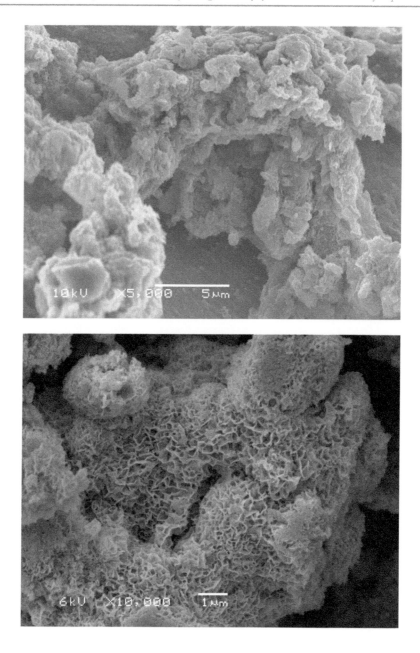

Figure: *Scanning Electron Microscopy (SEM) picture of biologically produced Birnessite (top), and SEM picture of physico-chemically produced Birnessite (bottom)*
(photos: Arie Zwijnenburg, Wetsus)

4 BIOLOGICAL AND PHYSICO-CHEMICAL FORMATION OF BIRNESSITE DURING RIPENING OF MANGANESE REMOVAL FILTERS

Main part of this chapter was published as:
Jantinus H. Bruins, Branislav Petrusevski, Yness M. Slokar, Koen Huysman, Koen Joris, Joop C. Kruithof, Maria D. Kennedy(2015). Biological and physicochemical formation of Birnessite during the ripening of manganese removal filters. Water Research, 69C: 154-161

4.1 Abstract

The efficiency of manganese removal in conventional groundwater treatment consisting of aeration followed by rapid sand filtration, strongly depends on the ability of filter media to promote auto-catalytic adsorption of dissolved manganese and its subsequent oxidation. Earlier studies have shown that the compound responsible for the auto-catalytic activity in ripened filters is a manganese oxide called Birnessite. The aim of this study was to determine if the ripening of manganese removal filters and the formation of Birnessite on virgin sand is initiated biologically or physico-chemically. The ripening of virgin filter media in a pilot filter column fed by pre-treated manganese containing groundwater was studied for approximately 600 days. Samples of filter media were taken at regular time intervals, and the manganese oxides formed in the coating were analysed by Raman spectroscopy, Electron Paramagnetic Resonance (EPR) and Scanning Electron Microscopy (SEM). From the EPR analyses, it was established that the formation of Birnessite was most likely initiated via biological activity. With the progress of filter ripening and development of the coating, Birnessite formation became predominantly physico-chemical, although biological manganese oxidation continued to contribute to the overall manganese removal. The knowledge that manganese removal in conventional groundwater treatment is initiated biologically could be of help in reducing typically long ripening times by creating conditions that are favourable for the growth of manganese oxidizing bacteria.

Keywords: Manganese removal; Ripening time; Manganese oxides; Birnessite; Biological manganese oxidation; Physico-chemical manganese oxidation

4.2 Introduction

As mentioned before, in Europe, the removal of manganese from groundwater for water supply is commonly achieved through conventional aeration-rapid sand filtration. This treatment is not only cost effective but also environmentally friendly, because no chemicals ($KMnO_4$, O_3, Cl_2) are required for oxidation of Mn^{2+}. The efficiency of manganese removal in aeration-filtration treatment of groundwater, and particularly during ripening of virgin filter media, strongly depends on the ability of filter media to adsorb dissolved manganese. Virgin quartz sand and anthracite, the most commonly used filter media, do not have significant capacity to adsorb dissolved manganese. According to the redox potential - pH diagram for aqueous manganese, chemical formation of MnO_x requires a high redox potential and pH (Stumm and Morgan, 1996). In general, in rapid sand filters used for groundwater treatment, conditions for chemical formation of auto-catalytically active manganese oxides on virgin filter media, which can adsorb and subsequently oxidize dissolved manganese, are poor since most groundwaters have a low redox potential and pH.

MnO_x-coated filter media could promote manganese removal through physico-chemical auto-catalytical adsorption and subsequent oxidation of adsorbed manganese, as described by the oxidation kinetics for dissolved Mn^{2+} by oxygen in aqueous solution (Stumm and Morgan, 1996):

$-d[Mn^{2+}]/dt$	$=$	$-k_0.[Mn^{2+}] + k_1.[Mn^{2+}].[MnO_X]$	(1)
k_0	$=$	$k.P_{O_2}.[OH^-]^2$	(2)

Where:

k_0	$=$	reaction rate constant (L/min)
k_1	$=$	reaction rate constant (L/mol.min)
P_{O_2}	$=$	partial pressure of oxygen (atm)

Equation (1) shows that physico-chemical Mn^{2+} oxidation is enhanced by heterogeneous autocatalytic activity of solid manganese oxides (*e.g.*, MnO_X) present in filter media coating. Furthermore equation (2) shows that the reaction rate constant k_0 is influenced by the oxygen concentration, and is also strongly pH dependent. On the other hand, biological formation of manganese oxide is less pH dependent. Burger *et.al.* (2008b) reported biological manganese removal at a pH of 6.5, whereas Hoyland *et. al.* (2014) found biological manganese removal at a pH as low as 6.3. Several researchers investigated the influence of different types of bacteria on manganese oxidation, amongst others, *Leptothrix* sp. (Adams and Ghiorse, 1985; Barger *et al.*, 2009; Boogerd and De Vrind, 1987; Burger *et al.*, 2008a, 2008b; Corstjens *et al.*, 1997; El Gheriany *et al.*, 2009; Hope and Bott, 2004; Tebo *et al.*, 2004, 2005), *Pseudomonas* sp. (Barger *et al.*, 2009; Brouwers *et al.*, 1999; Caspi *et al.*, 1998; DePalma, 1993; Gounot, 1994; Tebo *et al.*, 2004, 2005;, Villalobos *et al.*, 2003, 2006) and *Bacillus* sp. (Barger *et al.*, 2005, 2009; Brouwers *et al.*, 2000; Mann *et al.*, 1988; Tebo *et al.*, 2004, 2005). Furthermore several studies suggested that the manganese removal in rapid sand filters could be accelerated by the use of 'bio'-aged or coated filter media (Buamah *et al.*, 2008, 2009a, 2009b; Hu *et al.*, 2004a, 2004b; Islam *et al.*, 2010; Katsoyiannis and Zouboulis, 2004; Kim *et al.*, 2009; Sahabi *et al.*, 2009; Tiwari *et al.*, 2007). Therefore literature strongly suggests the impact of biology on manganese removal. Bruins *et al.* (2014a) established that the manganese oxide responsible for the autocatalytic action in ripened manganese removal filters is a mineral called Birnessite. Despite assumptions proposed in several literature references that manganese removal is a biological process (Burger *et al.*, 2008b; Vandenabeele *et al.*, 1992; Tekerlekopoulou *et al.*, 2008), Bruins *et al.* (2014a) showed that Birnessite present in coatings of ripened manganese removal filters (in operation for over 15 years) was of physico-chemical origin. However, their results do not exclude that initially Birnessite layers were formed biologically.

The main focus of this research was to establish how the filter media ripening process, regarding manganese removal in groundwater treatment based on aeration and rapid sand filtration was initiated. The objective of the study was to show whether the MnO_x (Birnessite) formed during the filter ripening was of biological and/or physico-chemical origin.

4.3 Materials and Methods

The ripening process was studied in a pilot filter column (A1) located at the full-scale groundwater treatment plant (GWTP) Grobbendonk, water supply company Pidpa, Belgium.

At GWTP Grobbendonk groundwater is treated containing iron, manganese and ammonia. The full-scale groundwater treatment plant at this location consists of (cascade) aeration, filtration, post aeration (cascade) with pH correction and post filtration. In the first filtration stage, iron is highly efficiently removed, while partial ammonia removal is achieved. In the second filtration stage the remaining ammonia and most of the manganese is removed. Filter media ripening of post filters with respect to manganese removal at this location is a very fast process, and is typically completed within 14 to 21 days. To simulate post filtration of the full scale GWTP, the pilot filter column was fed with water after post aeration with pH correction.

The filter column (A1) used in this study had an inner diameter of 10 cm and was filled with virgin quartz sand of size fraction 0.7-1.25 mm, with a height of 0.30 m. The filter column was operated in down flow mode at a filtration rate of 5.1 m/h with an empty bed contact time of 3.5 minutes. Backwashing was carried out after approximately every 2 weeks of continuous operation. Backwashing was performed with water only, at a backwash rate of 35 m/h, resulting in a filter bed expansion of approximately 20%.

To evaluate the ripening process of the filter media with respect to manganese removal and the formation of MnO_x, both water (feed, filtrate and backwash water), and filter media samples were collected at certain time intervals. Filter media samples were taken from the top of the filter bed. Samples of backwash water were allowed to settle, and settled particles were dried at room temperature.

To determine the pathway of Birnessite formation on the virgin sand during the ripening process and on the coated media after prolonged filter runs, and to characterize, identify and classify the manganese oxide in filter media coating, the following analytical methods were used:

- Raman spectroscopy;
- Electron paramagnetic resonance (EPR);
- Scanning electron microscopy (SEM).

The techniques, instruments and sample preparation methods applied are described by Bruins *et al.* (2014a). The obtained results were compared with references from literature and with a mineral Birnessite, originating from the Aufgeklärt Glück Mine Hasserode near Wernigerode, Harz, Saxony-Anhalt, Germany (Witzke, 2012).

The performance of the pilot filter was monitored for a filter run of 20 months. In this paper, the time needed to achieve very effective (> 95 %) manganese removal starting with virgin sand is called the 'filter ripening time'.

4.4 Results and Discussion

4.4.1 Analytical data pilot

Quality of feed water for the pilot column is given in Table 4.1.

Table 4.1: *Quality of feed water to the pilot filter column*

Parameter	unit	range
Manganese	mg/L	0.100 – 0.150
Iron[*]	mg/L	0.03 - 0.1
Ammonium	mg/L	0 - 0.20
pH	-	7.5 - 7.9[**]
Oxygen	mg/L	8 - 9.5
Turbidity	FTU	0.5 - 1.2
Redox potential	mV	+200 - +290
Temperature	°C	10.5 - 12.5
Hydrogen carbonate	mg/L	115 - 135

[*] *mainly in oxidized* (Fe^{3+}) *form*
[**] *most of the time pH ranged from 7.5 to 7.6*

From Table 4.1 it is clear that during this pilot study water quality conditions (pH and redox potential) enabled Mn^{2+} adsorption and subsequent oxidation on existing manganese oxide layers (Bruins *et al.*, 2014b). However it was doubtful if these water quality conditions were suitable to achieve a fast and effective start-up of the filter ripening process by adsorption and subsequent oxidation on virgin filter media (Graveland, 1971; Stumm and Morgan 1996).

Fig. 4.1 plots acidity and redox potential of feed water from this study (black dot), in the redox-pH diagram of aqueous manganese (Stumm and Morgan, 1996). Redox-pH conditions for aerated groundwater from a number of selected Dutch full-scale GWTPs are also included in the graph (grey dots).

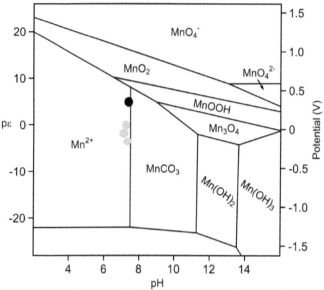

Figure 4.1: *Electron activity (pε) or redox potential (E_h in V) – pH diagram for aqueous manganese (adopted from Stumm and Morgan, 1996), compared to feed water conditions of the pilot (black dot) and groundwater of other GWTPs (grey dots).*

From Fig. 4.1 it can be seen that the pH and redox potential of feed water to the pilot (black dot) is in the transition zone of conditions for dissolved manganese and conditions theoretically required to form $MnCO_3$. At a given pH the amount of dissolved Mn^{2+} in groundwater is controlled by the concentration of hydrogen carbonate present in the water (Buamah, 2009b). Given the composition of the feed water to the pilot filter column (Table 4.1), significant formation of manganese carbonate was not likely. This assumption was confirmed by Raman spectroscopy analyses, by which no or negligible amounts of $MnCO_3$ were found in the coating. However, $MnCO_3$ particles could be formed in limited amounts in the pilot feed water, after pH correction (dosing of milk of lime, applied at the full-scale GWTP Grobbendonk upstream of the intake point for the pilot, with temporarily locally a high pH). Consequently, the presence of small amounts of $MnCO_3$ particles deposited on the (virgin) filter media in the filter column of the pilot cannot be excluded. For the ripening process this might be a disadvantage, because $MnCO_3$ has no auto-catalytical adsorption and oxidation properties, unlike manganese oxides (MnO_X). Furthermore, $MnCO_3$ precipitate can cover already formed (auto-catalytically active) MnO_X (Graveland, 1971; Graveland and Heertjes, 1975). Hence, in this way $MnCO_3$ is able to hinder the ripening process by retarding and disturbing the auto-catalytical adsorption and oxidation reaction. Consequently, possible formation of $MnCO_3$ and its retention on filter media will not be beneficial for the process of filter media ripening. During this study, however, the negative effect of $MnCO_3$ was limited, because in line with the $MnCO_3$ solubility (Buamah, 2009b) no or negligible amounts of $MnCO_3$ were identified with Raman spectroscopy in the coating.

4.4.2 Ripening time of filter media

Fig. 4.2 gives an overview of the manganese removal efficiency during the first 25 days of the ripening process of the pilot column. From Fig. 4.2 it can be seen that ripening of the filter media (sand) was completed after approximately 25 days.

A ripening period of approximately 25 days is relatively short. Although in some situations low Mn^{2+} concentration may lead to a relatively short ripening time, in general, it typically takes more than two to three months, and in some exceptional cases even more than a year, to achieve complete manganese removal even with low Mn^{2+} concentrations (Bahlman, 2014; Cools, 2010; Krull, 2010).

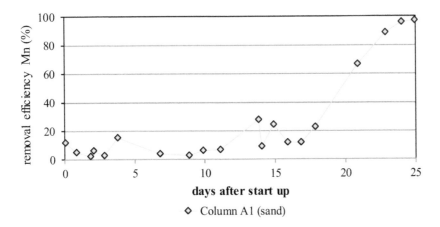

Figure 4.2: *Ripening period of pilot column A1 (height 0.3 m, diameter 0.1 m), filtration rate 5.1 m/h.*

The build-up of MnO_X layers on virgin filter media can be seen using a microscope. Fig. 4.3 shows images of virgin sand and the same filter media after a filter run time of 81 days. The observed thin grey layer with (tiny) blackish spots (Fig. 4.3 (R)) represents MnOx deposits.

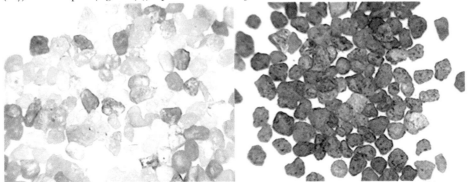

Figure 4.3: *Virgin sand (L) and sand from column A1 after a filter run time of 81 days (R). Microscope, magnification 20x.*

4.4.3 Raman spectroscopy

Raman spectra of a mineral Birnessite reference, MnO_X deposits on filter media (from pilot column A1) 18 days after initiating filter media ripening, and MnO_X in solids from the backwash water (BW) from the pilot column (28 days after initiation of filter media ripening) were compared. From Fig. 4.2 it can be seen that the sand sample from the pilot column was taken at the stage of exponential increase of manganese removal efficiency. A few hours before sampling the coated sand, the manganese removal efficiency was approximately 25%, while two days later the manganese removal efficiency was increased to approximately 65% (Fig. 4.2). Consequently, the sand sample was taken at a very early stage in the ripening process at which the first manganese oxides were formed.

From Raman spectra comparison it was found that the main peaks obtained for solids from backwash water and sand from the pilot column were comparable with the Raman shift, of the main peaks of the Birnessite reference (Witzke, 2012). The positions of the peaks found in this study were also in accordance with the Birnessite peaks found in literature (Ma *et al.*, 2007). Consequently, this strongly suggests that manganese oxide formed on the filter media already at the early stage in the ripening process was Birnessite.

After a number of filter run times Raman spectra were analysed for several filter media samples and solids from backwash water (Fig. 4.4). For these samples the main peaks were always found at comparable Raman shifts. From this it can be concluded that during the complete filter run time, the manganese oxide found in the filter media coating and the solids in the backwash water was Birnessite.

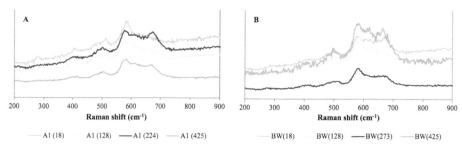

Figure 4.4: *Raman spectra of manganese oxide for different filter media samples from the pilot column (A1) and different samples of solids from backwash water (BW), both taken at different filter run times.*

Observed differences in the height of peaks for different samples (Fig. 4.4) can likely be attributed to different concentrations of counter ions (*e.g.*, sodium, calcium, etc.), present in the lattice of the manganese oxide, as reported by Julien *et al.* (2003, 2004). The presence of other ions was confirmed by ICP-MS and SEM-EDX analysis. Although the chemical formula(s) of Mn-oxides belonging to the Birnessite group can be expressed in different ways, they always include several counter ions (*e.g.*, $(Na^+,Ca^{2+},Mn^{2+})Mn_7O_{14}$ $2.8H_2O$, Post, 1999).

From the Raman spectroscopy analyses it can be concluded that manganese oxide in the filter media coating, formed during the ripening, and the solids present in the filter backwash water, were a Birnessite type of manganese oxide. In addition it can be concluded that Birnessite was already present at a very early stage of the filter ripening process. Finally, the peaks in the Raman spectra of both the filter media coating and solids from the backwash water were not very sharp, indicating that the Birnessite is poorly crystalline or amorphous.

4.4.4 Electron paramagnetic resonance (EPR)

Kim *et al.* (2011) recently reported that electron paramagnetic resonance (EPR) spectra could distinguish biologically formed Birnessite from inorganic (abiogenic) Birnessite. Especially the measured wavelength, expressed in gauss, was reported to be a criterion to distinguish between the two sources of Birnessite. Kim *et al.* (2011) made the following classification based on wavelength (ΔH) in gauss (measured at room temperature - 290 K):

Birnessite of bacterial origin	: ΔH	< 600	gauss
Birnessite of biomineral origin *)	: ΔH	600-1200	gauss
Birnessite of abiogenic origin	: ΔH	> 1200	gauss

)biomineral = a mineral (Mn nodules from fresh and marine water and desert varnish) of possible biological origin, that has been transformed over a long time into an abiogenic mineral.

Fig. 4.5 shows the EPR spectrum of the Birnessite reference (Witzke, 2012) used in this study. The calculated wavelength (ΔH) obtained from this Birnessite reference is significantly larger than 1200 G, confirming its abiogenic (inorganic) origin.

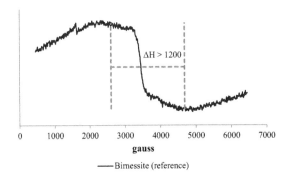

——Birnessite (reference)

Figure 4.5: *EPR spectrum at 290 K of the Birnessite reference.*

EPR spectra were also measured for the collected samples of (coated) sand from the filter column and samples of solids from the backwash water. In Fig. 4.6 the calculated wavelengths (ΔH in gauss) of the EPR spectra of coated sand samples (red diamonds), and the backwash water samples (green triangles) were plotted as a function of the filter run time (in days). The blue dot in Fig. 4.6 shows the ΔH value of Birnessite present in the filter media coating of a full-scale filter of the 2nd filtration stage at GWTP Grobbendonk, after a filter run of at least 4 years. This full-scale filter received the same feed water as the pilot column.

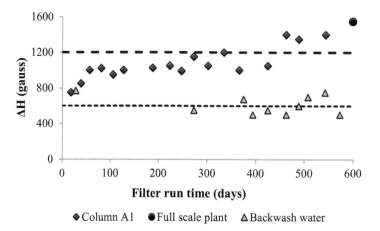

◆ Column A1   ● Full scale plant   △ Backwash water

Figure 4.6: *Wavelengths (ΔH in gauss) of EPR Birnessite spectra from coated sand samples from the pilot column (red diamonds), and solid samples from the backwash water (green triangles), versus filter run time.*

From Fig. 4.6 it can be seen that the ΔH values of the Birnessite present in the coating of filter media increased with the filter run time. ΔH values at the start of the filter operation were just above the fine dotted line representing the boundary between Birnessite of biological and biomineral origin. In this study, values in the biomineral range might indicate a mixture of biological and abiogenic Birnessite. With a prolonged filter run time, ΔH values increased up to values above the coarse dotted line, reflecting the presence of Birnessite of physico-chemical origin. The increase in ΔH can be explained by the physico-chemical manganese adsorption/oxidation reaction due to the presence of formed manganese oxides on the media surface. It is known that the physico-chemical manganese adsorption/oxidation reaction is an extremely fast auto-catalytic process (Graveland, 1971; Stumm and Morgan, 1996). Hence, the amount of physico-chemically formed Birnessite on the filter media may increase compared to the amount of biologically formed Birnessite.

A rapid increase in ΔH values was observed initially. Given the exponential shape of the ΔH curve, and the fact that the first sample was taken 18 days after the start of the filter run, it is very likely that the initial values of ΔH were even below the value of 600 G proposed by Kim *et al.* (2011) for Birnessite of biological origin. Samples of material from the coating taken after a prolonged filter run time showed ΔH values approaching that of Birnessite of abiogenic origin. Samples of a coating from an aged full-scale filter showed a ΔH value representing Birnessite of abiogenic origin (ΔH > 1200 G).

The average ΔH value of Birnessite present in solid samples collected from the backwash water was 609 G, with a standard deviation of 111 G. This suggests that Birnessite found in filter backwash water was predominantly of biological origin. ΔH values of Birnessite found on solids in backwash water were consistently lower than values found for Birnessite present in the filter media coating throughout the whole testing period of approximately 600 days.

EPR results suggest that physico-chemical formation of manganese oxide on filter media is the predominant process, once initial layers of biological Birnessite are formed. At the same time, results of the EPR analysis of solids found in the filter backwash water indicate that biological oxidation of Mn^{2+} takes place throughout the whole filter run time. Biologically formed Birnessite in solids found in backwash water are likely formed on and around the bacterial cells attached on the surface of the filter media. This biomass present on filter grains with formed Birnessite solids is partially flushed out during backwashing. Backwashing of the pilot column is done with water only, with filter bed expansion ≤20%, which is not very abrasive, and hence, it can be assumed that only a limited amount of Birnessite of physico-chemical origin, that is strongly attached to filter media, will be flushed out.

Results emerging from this EPR study strongly suggest that filter media ripening with respect to manganese removal starts biologically, and at a later stage is controlled by the physico-chemical oxidation of adsorbed Mn^{2+}.

The conclusions of this study based on the manganese oxide identification and characterization by Raman spectroscopy, EPR and SEM can be further underpinned by showing the presence of manganese oxidizing bacteria. Therefore molecular techniques such as qPCR, Malditof and next generation DNA sequencing were applied. With these techniques the presence of bacteria generally known to be able to oxidize Mn^{2+} such as *Pseudomonas* sp. and *Leptothrix* sp. were confirmed. These results confirm the conclusions based on Raman spectroscopy, EPR and SEM measurements of manganese oxides in filter coatings described in this paper. The detailed identification of manganese oxidizing bacteria and their relevance for manganese removal in practice is a comprehensive study, which will be published in a separate paper (Bruins *et al.*, to be published in 2015).

4.4.5 Scanning electron microscopy (SEM)

To examine the structural difference between biologically and physico-chemically formed Birnessite, samples of filter media taken after different filter run times were analysed by SEM. Jiang *et al.* (2010) showed a distinctive difference in structure between biologically (*Pseudomonas putida*) and physico-chemically formed Birnessite by SEM imaging. Biologically produced Birnessite has a fluffy plate structure in contrast to physico-chemically formed Birnessite that has a more coral or sponge type structure. The SEM images of filter media coating from this study are compared with the findings of Jiang *et al.* (2010).

Fig. 4.7 shows SEM micrographs of Birnessite found in the coating of the filter media after different run times.

Figure4.7: *SEM image of Birnessite found in the coating of the filter media (sand) after 128 days (left), 394 days (middle) and 273 days (right) of filter run time (magnification resp. 5,000x; 5,000x; 10,000x).*

Fig. 4.7 (left) shows the presence of fluffy plate structures, typical for Birnessite of biological origin (Jiang *et al.*, 2010). The SEM micrograph of aged filter media (Fig. 4.7, middle), shows coral or sponge-like structures, typical for Birnessite of physico-chemical origin (Jiang *et al.*, 2010).

Fig. 4.7 (right) shows a SEM micrograph of Birnessite present in aged filter media made at a larger magnification (10,000x). Structures typical for physico-chemically formed Birnessite are clearly visible *e.g.,* 'layered' patterns within the coral or sponge structure.

According to Post (1999), Birnessite consists of stacked octahedral MnO_6 layers, with different interlayer cations and water. According to Post (1999) such a structure, allows Birnessite to 'readily participate in oxidation-reduction and cation-exchange reactions, and therefore plays a significant role in soil and groundwater chemistry'. This consequently makes Birnessite extremely suitable for the auto-catalytic oxidation reaction in conventional aeration-filtration GWTPs.

The SEM images also confirmed that the Birnessite is poorly crystalline or amorphous.

The character of manganese removal being biological or physico-chemical has been debated for quite some time. Some researchers propose a physico-chemically driven process (Graveland and Heertjes, 1975), whereas others suggest a biological process (Tekerlekopoulou *et al.*, 2008; Burger *et al.*, 2008a, Vandenabeele *et al.*, 1992). When strong oxidizers, such as chlorine, ozone and potassium permanganate are used, it is obvious that the formed MnO_2 (Pyrolusite) is produced in a physico-chemical way (Knocke *et al.*, 1991). However, when groundwater treatment by conventional aeration-filtration, without the use of strong oxidants, is applied the common quality of groundwater (relatively low pH, low redox potential) is, in general, not supportive for a fast start of virgin filter media ripening through a physico-chemical mechanism (Bruins *et al.*, 2014b). Burger *et.al.*(2008b) and Hoyland *et. al.* (2014) have shown that contrary to chemical oxidation, biological oxidation of Mn^{2+} can take place at low pH. The current study shows that biological Mn^{2+} oxidation plays a decisive role in filter media ripening. After prolonged time the formed manganese oxide Birnessite becomes of a more physico-chemical origin, although part of the Birnessite still has a biological origin. However, results emerging from this study were insufficient to clarify if the biological oxidation of manganese remains the controlling manganese removal mechanism also after initial ripening of the filter media. Although substantial research has been performed on this topic, little is known why heterotrophic bacteria oxidize manganese and suggested reasons are at least speculative (De Schamphelaire *et al.*, 2007). Therefore, it is important to unravel how heterotrophic bacteria (*e.g., Pseudomonas* sp., *Leptothrix* sp.), also found in this research, can benefit from this oxidation reaction. Furthermore the question must be answered if *Pseudomonas* sp. and *Leptothrix* sp. are acting alone or in cooperation with other bacteria. Understanding the biochemical and physico-chemical mechanisms by which manganese coating of filter media is formed during the filter ripening could help creating conditions favouring Birnessite formation. This knowledge together with the results of this study may help reducing long ripening periods of manganese removal filters with virgin filter media, making groundwater treatment by conventional aeration-filtration more generally applicable.

4.5 Conclusions

The following conclusions can be drawn based on the results emerging from this study:

- Analysis of the filter media from the manganese removal plant conducted with Raman spectroscopy, Electron Paramagnetic Resonance (EPR) and Scanning Electron Microscopy (SEM) showed that a Birnessite type of manganese oxide is the predominant mineral in the coating;
- Raman spectroscopy results showed that Birnessite is already present in the coating at a very early stage of the ripening process;

- From the EPR analyses and comparison with literature, the Birnessite type of manganese oxide, at the beginning of the ripening process was most likely of biological origin. Over a prolonged filter run time, Birnessite changed from a predominantly biologically formed to a physico-chemically formed compound;

- Solids collected from filter backwash water throughout the whole ripening period were consistently of biological origin, suggesting that biological oxidation of adsorbed manganese was present throughout the filter run, and contributed to manganese removal;

- SEM micrographs showed a clear difference between biologically and physico-chemically formed Birnessite. Biologically produced Birnessite is fluffy, plate structured, whereas physico-chemically produced Birnessite shows more a sponge or coral structure;

- Understanding the mechanisms by which a manganese coating of filter media starts up could endorse the creation of conditions favouring Birnessite formation, and possibly help in reducing typically long ripening periods of manganese removal filters with virgin filter media.

4.6 Acknowledgements

This research was financially and technically supported by WLN and the Dutch water supply companies Groningen (WBG) and Drenthe (WMD). The authors are also grateful to Dr. Arie Zwijnenburg and Mr. Ton van der Zande (Wetsus) for providing the SEM and Raman spectroscopy analyses and to Dr. Wesley Browne (University of Groningen) for giving access to the Electron Paramagnetic Resonance measurements. Also thanks to Dr. Witzke for providing a mineral Birnessite (reference) sample. Last but certainly not least, thanks go to the water company Pidpa (Belgium) for their willingness to share the data from their groundwater treatment plant with us and to give us the opportunity to perform a pilot test at Grobbendonk. We especially thank Mrs. Ann Maeyninckx and Mrs. Martine Cuypers (Pidpa) for fulfilling a great task by running the pilot and performing in situ analyses.

4.7 References

Adams, L.F., Ghiorse, W.C., 1985. Influence of Manganese on Growth of a Sheathless Strain of *Leptothrix discophora*. Applied and Environmental Microbiology, 49 (3), 556 –562.

Bahlman, J.A., 2014. Evides Waterbedrijf (Dutch water company), personal communication, The Netherlands.

Barger, J.R., Tebo, B.M., Bergmann, U., Webb, S.M., Glatzel, P., Chiu, V.Q., Villalobos, M., 2005. Biotic and abiotic products of Mn (II) oxidation by spores of the marine *Bacillus* sp. strain SG-1. American Mineralogist, 90, 143-154.

Barger, J.R., Fuller, C.C., Marcu, M.A., Brearly, A., Perez De la Rosa M., Webb, S.M., Caldwell, W.A., 2009. Structural characterization of terrestrial microbial Mn oxides from Pinal Ckeek, AZ. Ceochimica et Cosmochimica Acta 73, 889-910.

Boogerd, F.C., De Vrind, J.P.M., 1987. Manganese oxidation by *Leptothrix discophora*. Journal of Bacteriology, 489-494.

Brouwers, G.J., De Vrind, J.P.M, Corstjens, P.L.A.M., Cornelis, P., Baysse, C., De Vrind-De Jong, E.W., 1999. CumA, a Gene Encoding a multicopper oxidase, is involved in Mn^{2+} oxidation in *Pseudomonas putida* GB-1. Applied and Environmental Microbiology, 65(4), 1762-1768.

Brouwers, G.J., Vijgenboom, E., Corstjens, P.L.A.M., De Vrind, J.P.M., De Vrind-De Jong, E.W., 2000. Bacterial Mn2+ oxidizing systems and multicopper oxidases: an overview of Mechanisms and Functions. Geomicrobiology Journal, 17(1), 1-24.

Bruins, J.H., Petrusevski, B., Slokar, Y.M., Kruithof, J.C., Kennedy, M.D., 2014a. Manganese removal from groundwater: characterization of filter media coating. Desalination and Water Treatment (in press, on line 18 June 2014, DOI:10.1080/19443994.2014.927802).

Bruins, J.H., Vries, D., Petrusevski, B., Slokar, Y.M., Kennedy, M.D., 2014b. Assessment of manganese removal from over 100 groundwater treatment plants. Journal of Water Supply: Research and Technology-AQUA, 63.4 268-280.

Buamah, R., Petrusevski, B., Schippers, J.C., 2008. Adsorptive removal of manganese (II) from the aqueous phase using iron oxide coated sand. Journal of Water Supply: Research and Technology-AQUA 57.1, 1-11.

Buamah, R., Petrusevski, B., de Ridder, D., Van de Watering and Schippers, J.C., 2009a. Manganese removal in groundwater treatment: practice, problems and probable solutions. Journal of Water Science and Technology: Water Supply 9.1: 89 - 98.

Buamah, R., 2009b. Adsorptive removal of manganese, arsenic and iron from groundwater. PhD thesis, Unesco-IHE, Delft, Wageningen University, The Netherlands.

Burger, M.S., Krentz, C.A., Mercer, S.S., Gagnon, G.A., 2008a. Manganese removal and occurrence of manganese oxidizing bacteria in full-scale biofilters. Journal of Water Supply: Research and Technology-AQUA 57.5.

Burger, M.S., Mercer, S.S., Shupe, G.D., Gagnon, G.A., 2008b. Manganese removal during bench-scale biofiltration, Water Research 42, 4733-4742.

Caspi, R., Tebo, B.M., Haygood, M.G., 1998. C-Type Cytochromes and Manganese Oxidation in *Pseudomonas putida* MnB1. Applied and Environmental Microbiology, 64 (10), 3549-3555.

Cools, B., 2010. De Watergroep (Flemish water company), personal communication, Belgium.

Corstjens, P.L.A.M., De Vrind, J.P.M., Goosen, T., De Vrind-de Jong, E.W., 1997. Identification and molecular analysis of the Leptothrix discophora SS-1 mofA gene, a gene putatively encoding a manganese-oxidizing protein with copper domains. Geomicrobiology Journal, 14 (2), 91-108.

DePalma, S.R., 1993. Manganese oxidation by *Pseudomonas putida*. PhD thesis, Harvard University, Cambridge, USA.

De Schamphelaire, L. , Rabaey, K. , Boon, N. , Verstraete, W. And Boeckx, P. 2007. Minireview: The potential of enhanced manganese redox cycling for sediment oxidation. Geomicrobiology Journal, 24 (7), 547-558.

El Gheriany, I.A., Bocioaga, D., Hay A.G, Ghiorse, W.C., Shuler, M.L., Lion, L.W., 2009. Iron Requirement for Mn (II) Oxidation by *Leptothrix discophora* SS-1. Applied and Environmental Microbiology, 75 (5), 1229-1235.

Gounot, A-M., 1994. Microbial oxidation and reduction of manganese: Consequences in groundwater and applications. FEMS Microbiology Reviews 14, 339-350.

Graveland, A., 1971. Removal of manganese from groundwater. PhD thesis, Technical University Delft, The Netherlands.

Graveland, A., Heertjes, P.M., 1975. Removal of manganese from groundwater by heterogeneous autocatalytic oxidation. Trans. Instn. Chem. Engrs., 53, 154-164.

Hope, C.K., Bott, T.R., 2004. Laboratory modelling of manganese biofiltration using biofilms of *Leptothrix doscophora*, Water Research 38, 1853-1861.

Hoyland, V.W., Knocke, W.R., Falkinham III, J.O., Prude, A., Singh, G., 2014. Effect of drinking water process treatment parameters on biological removal of manganese from surface water, Water Research, in press (on line since Augustus 2014).

Hu, P-Y., Hsieh, Y-H., Chen, J-C., Chang, C-Y., 2004a. Adsorption of divalent manganese ion on manganese-coated sand. Journal of Water Supply: Research and Technology-AQUA 53.3, 151-158.

Hu, P-Y., Hsieh, Y-H., Chen, J-C., Chang, C-Y., 2004b. Characteristics of manganese-coated sand using SEM and EDAX analysis. Journal of Colloid and Interface Science, 272, 308-313.

Islam, A.A., Goodwill, J.E., Bouchard, R., Tobiasen, J.E., Knocke, W.R., 2010. Characterization of filter media $MnO_2(s)$ surfaces and Mn removal capability. Journal AWWA, 102 (9), 71-83.

Jiang, S., Kim, D-G., Kim, J., Ko, S-O., 2010. Characterization of the biogenic manganese oxides produced by *Pseudomona putida* strain MnB1. Environ. Eng. Res., 15 (4), 183-190.

Julien, C., Massot, M., Baddour-Hadjean, R., Franger, S., Bach, S., Pereira-Ramos, J.P., 2003. Raman spectra of birnessite manganese dioxides. Solid State Ionics 159, 345– 356.

Julien, C., Massot, M., Poinsignon, C., 2004. Lattice vibrations of manganese oxides Part I. Periodic structures. Spectrochimica Acta Part A 60, 689-700.

Katsoyiannis, I. A., Zouboulis A.I., 2004. Biological treatment of Mn (II) and Fe (II) containing groundwater: kinetic considerations and product characterization. Water Research 38, 1922-1932.

Kim, W.G., Kim, S.J., Lee, S.M., Tiwari, D., 2009. Removal characteristics of manganese-coated solid samples for Mn (II). Desalination and Water Treatment 4, 218-223.

Kim, S.S., Bargar, J.R., Nealson, K.H., Flood, B.E., Kirschvink, J.L., Raub, T.D., Tebo, B.M., Villalobos, M., 2011. Searching for Biosignatures Using Electron Paramagnetic Resonance (EPR) Analysis of Manganese Oxides. Astrobiology, 11(8), 775-786.

Knocke, W.R., Van Benschoten, J.E., Kearny, M.J., Soborski, A.W. & Reckhow, D.A. 1991. Kinetics of Manganese and Iron oxidation by Potassium Permanganate and Chlorine dioxide. *Journal of AWWA*, June, 80-87.

Krull, J., 2010. Stadtwerke Emden (German water company), personal communication, Germany.
Ma, S-B., Ahn, K-Y., Lee, E-S., Oh, K-H., Kim, K-B., 2007. Synthesis and characterization of manganese dioxide spontaneously coated on carbon nanotubes. Carbon 45, 375-382.

Mann, S., Sparks. N.H.C., Scoot, G.H.E., De Vrind-De Jong E.W., 1988. Oxidation of Manganese and formation of Mn_3O_4 (Hausmannite) by spore coats of a marine *Bacillus* sp. Environmental Microbiology, 54 (8), 2140-2143.

Post, J. E., 1999. Manganese oxide minerals: Crystal structures and economic and environmental significance. Proceedings of the National Academy of Sciences USA, 96, 3447-3454.

Sahabi, D.M., Takeda, M., Suzuki, I., Koizumi, J-I., 2009. Removal of Mn^{2+} from water by "aged" biofilter media: The role of catalytic oxides layers. Journal of Bioscience and Bioengineering, 107(2), 151-157.

Stumm, W. and Morgan, J.J., 1996. Aquatic Chemistry, chemical equilibria and rates, 3rd ed. Wiley, New York.

Tebo, B.M., Marger, J.R., Clement, B.G., Dick, G.J., Murray, K.J., Parker, D., Verity, R., Webb, S.M., 2004. Biogenic Manganese oxides: Properties and mechanisms of formation. Annu. Rev. Earth Planet Sci., 32, 287-328.

Tebo, B.M., Johnson, H.A., McCarthy, J.K., Templeton, A.S., 2005. Geomicrobiology of manganese (II) oxidation. TRENDS in Microbiology, 13(9), 421-428.

Tekerlekopoulou, A.G., Vasiliadou, I.A., Vayenas, D.V., 2008. Biological manganese removal from potable water using trickling filters. Biochemical Engineering Journal 38, 292-301.

Tiwari, D., Yu, M.R., Kim, M.N., Lee, S.M., Kwon, O.H., Choi, K.M., Lim, G.J., Yang, J.K., 2007. Potential application of manganese coated sand in the removal of Mn (II) from aqueous solutions. Water Science & Technology, 56 (7), 153-160.

Vandenabeele, J., De Beer, D., Germonpré, R., Verstreate, W., 1992. Manganese oxidation by microbial consortia from sand filters. Microbial Ecology 24, 91-108.

Villalobos, M., Toner, B., Bargar, J., Sposito, G., 2003. Characterization of the manganese oxide produced by *Pseudomonas putida* strain MnB1. Geochimica et Cosmochimica Acta, 67(14), 2649-2662.

Villalobos, M., Lanson, B., Manceau, A., Toner, B., Sposito, G., 2006. Structural model for the biogenic Mn oxide produced by *Pseudomonas putida*. American Mineralogist, 91, 489-502.

Witzke, T., 2012. Mineralogist PANanalytical, personal communication, Germany.

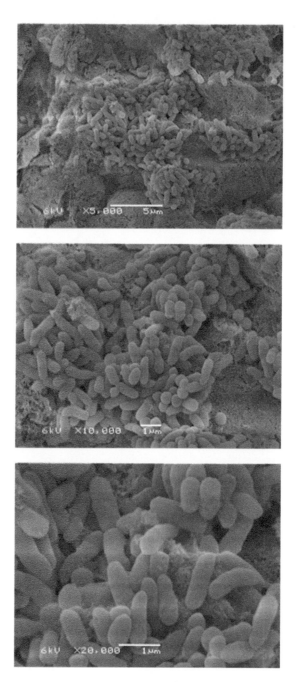

Figure: *Pseudomonas grimontii (from top to bottom; magnification 5,000x | 10,000x | 20,000x - Photos made by Jelmer Dijkstra (Wetsus), sample preparation by Pim Willemse (WLN).*

5 IDENTIFACTION OF THE BACTERIAL POPULATION IN MANGANESE REMOVAL FILTERS

Main part of this chapter was submitted as:

Jantinus H. Bruins, Branislav Petrusevski, Yness M. Slokar, Gerhard H. Wübbels, Koen Huysman, Bart A. Wullings, Koen Joris, Joop C. Kruithof, Maria D. Kennedy(2016). Identification of the bacterial population in manganese removal filters. Submitted to Water Science and Technology: Water Supply.

5.1 Abstract

Rapid ripening of manganese removal filters with virgin filter media, in conventional aeration-rapid sand filtration treatment of groundwater, greatly depends on the rate of formation of adsorptive MnO_x on filter media. Earlier studies have shown that the manganese oxide responsible for autocatalytic manganese removal in ripened filters is biologically formed Birnessite. The aim of this study was to identify bacteria present in freshly ripened filters for manganese removal. Samples of backwash water were taken from 1st stage (iron removal) filters and the freshly ripened manganese removal filter. The bacterial population was identified with 'next generation' DNA sequencing, and specific bacteria were quantified with qPCR and characterized by MALDI-TOF MS analysis.

The 'next generation' DNA sequencing analysis showed a bacteria population shift from the iron oxidizing species *Gallionella* sp. in the iron removal filter to manganese and nitrite oxidizing species *Pseudomonas* sp. and *Nitrospira* sp., respectively present in the manganese removal filter.

qPCR analysis confirmed the presence of a low concentration of the well-known Mn^{2+}-oxidizing species *P. putida* in the manganese removal filter backwash water.

Bacteria of the genus *Pseudomonas,* isolated from backwash water from the manganese removal filter were cultured and identified with MALDI-TOF MS analysis. Amongst others, *P. gessardii, P. grimontii, and P. koreensis* were identified.

The presence of high(er) concentrations of *Pseudomonas* spp. and minor amounts of *P. putida* and *Leptothrix* sp. in the ripened filter media supports the assumption that a microbial consortium is involved in the oxidation of manganese. However, inoculation of an isolated *Pseudomonas* species in a fermenter did not result in the production of MnOx under the performed laboratory conditions, whereas *P. putida* strain (ATCC 23483) was able to do so.

Keywords: biological manganese oxidation, Birnessite, manganese removal ripening time, molecular DNA techniques, Pseudomonas sp.

5.2 Introduction

An important drawback of aeration-rapid sand filtration, commonly applied in several West European countries to remove manganese from groundwater, is the long filter media ripening period. Farnsworth *et al.* (2012) reported that manganese oxide formed on the filter media, responsible for manganese removal, is a Birnessite type of mineral. Due to its structure, Birnessite has outstanding properties to adsorb and subsequently oxidize Mn^{2+} (Post and Veblen, 1990; Post 1999). Bruins *et al.* (2015a) showed that Birnessite present in the coating of a ripened manganese removing filter in operation for over 15 years, was of physicochemical origin. Chemical formation of Birnessite requires alkaline conditions (Feng *et al.,* 2005). The redox potential - pH diagram for aqueous manganese (Stumm and Morgan, 1996), suggest that besides a high pH, a high redox potential is required for chemical formation of MnO_x. Such water quality characteristics are not common for groundwater with, usually, low pH and low redox potential. Using electron paramagnetic resonance and Raman spectroscopy, Bruins *et al.* (2015b), showed that formation of Birnessite most likely starts through biological activity. In a number of other studies, it was also proposed that manganese removal is an obligatory biological process (Burger *et. al.*, 2008a,b, Farnsworth *et al.*, 2012; Katsoyiannis and Zouboulis, 2004; Vandenabeele *et al.*, 1992; Tekerlekopoulou *et al.*, 2008). Several species of bacteria (*Pseudomonas* sp., *Leptothrix* sp. and *Bacillus* spores) able to-, or involved in oxidation of Mn^{2+} have been identified in (ground)water (Tebo *et al.*, 2005; Kim *et. al.*, 2011). *Pseudomonas* sp., in particular *Pseudomonas putida*, were extensively studied in relation to manganese oxidation (Barger *et al.*, 2009; Brouwers *et al.*, 1999; Caspi *et al.*, 1998; DePalma, 1993; Gounot, 1994; Tebo *et al.*, 2004, 2005; Villalobos *et al.*, 2003, 2006). Similar studies were performed for *Leptothrix* sp. (Adams and Ghiorse, 1985; Barger et al., 2009; Boogerd and De Vrind, 1987; Burger *et al.*, 2008a, 2008b; Corstjens *et al.*, 1997; El Gheriany *et al.* 2009; Hope and Bott, 2004; Tebo *et al.* 2004, 2005) and *Bacillus* sp. spores (Barger *et al.*, 2005, 2009; Brouwers *et al.*, 2000; Mann *et al.*, 1988; Tebo *et al.*, 2004, 2005; Vrind de, *et al.*, 1986). Experiments with (spores of) *Bacillus* sp. were, however, often done with species of marine origin (Mann et al., 1988; Webb *et al.*, 2005a, 2005b).

Pseudomonas sp. and *Leptothrix* sp. are heterotrophic bacteria. *Leptothrix* sp. is able to oxidize iron as well as manganese, whereas *Pseudomonas* sp. is able to oxidize manganese only (Daum *et al.*, 1998; Fleming *et.al*, 2011). Despite extensive research, it is still not clear how heterotrophic bacteria benefit from manganese oxidation (De Schamphelaire *et al.*, 2007; Tebo *et al.* 2005; Geszvain *et al.*, 2013). However, from literature it is known that Mn^{2+} oxidation to MnO_2, by *Pseudomonas putida*, could be beneficial for the co-metabolic degradation of micro pollutants, resulting in the formation of easily accessible organic compounds for their metabolism (Sabirova *et al.*, 2008; Forrez *et al.*, 2010; Meerburg *et al.*, 2012). Similarly, it is proposed that complex organic molecules (*e.g.*, Natural Organic Matter (NOM), such as humic acids in groundwater) undergo degradation by the same process, performed by *e.g.*, *Pseudomonas putida* (Verstraete, 2013). Based on the proposed model of Mn^{2+} oxidation by *e.g.*, *Pseudomonas putida* and co-metabolic degradation of organic micro pollutants from Meerburg *et al.* (2012), a simplified degradation scheme for NOM is adopted (Fig. 5.1). The process of co-metabolic degradation it is titled "bio cracking".

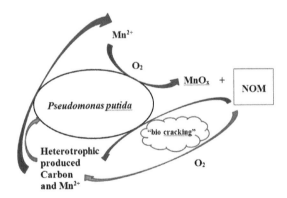

Figure 5.1: *Simplified co-metabolic degradation scheme of NOM by biological MnO_x oxidation (adopted from Meerburg et al.,(2012) and Verstraete (2013)).*

From Fig. 5.1 it can be seen that MnO_x, formed during the process of bio cracking, is reduced again to Mn^{2+}, in a self-supporting metabolic cycle. Literature suggested that not one bacterium is responsible for manganese oxidation, but a microbial consortia (Vandenabeele *et al.*, 1992; Vandenabeele, 1993; Verstraete, 2013). Once biological Birnessite (MnO_x) is formed on filter media, it has extremely high adsorptive capacity for metal ions, such as Mn^{2+} (Webb *et al*, 2005; Jiang *et al.*, 2010). In this way biologically produced Birnessite promotes and accelerates manganese removal through physicochemical autocatalytic adsorption and subsequent oxidation of adsorbed manganese according to the oxidation kinetics of dissolved Mn^{2+} by oxygen in aqueous solution (Stumm and Morgan, 1996).

The goal of this study was to provide additional insight in the role of biology in manganese removal. Specific objective of the study was to identify (with molecular (DNA) techniques) species of bacteria present in iron removal, and freshly ripened manganese removal filters. Furthermore the capability of selected bacterial species found in manganese removal filters to oxidize Mn^{2+} was investigated in the laboratory. This study will enhance knowledge of the role of biological activity in the ripening of manganese filters in practice and show how to create conditions favorable for the biological manganese oxidation process.

5.3 Materials and Methods

The experiments presented in this study were carried out on a pilot plant located at full scale groundwater treatment plant (GWTP) 'Grobbendonk', water supply company Pidpa, Belgium.

The full scale plant consists of a pre-aeration step (cascade), 1st filtration stage, post aeration (cascade) with pH correction, and post filtration. GWTP Grobbendonk treats groundwater containing iron, manganese and ammonia. Very effective removal of iron (> 98 %) is achieved in the 1st filtration step. Iron is removed predominantly through adsorptive and biological mechanisms, with support of the bacterium *Gallionella* sp. Filter media ripening of the post filters in this plant, concerning manganese removal, is a very fast process (typically complete manganese removal is achieved within approx. 16 days). It is believed that the filter media ripening process at this location is initial biologically. Similar to post filters in the full scale plant, the pilot filter column was fed with re-aerated filtrate from the first filter stage of the full scale GWTP, after pH correction (pH = 7.6). The pilot filter column is named in the rest of the paper as 'column A1'. Details of GWTP Grobbendonk and the pilot filter column are given in Bruins *et al.* (2015b).

To determine the presence and composition of the bacteria population, during the filter media ripening process samples were taken from backwash water from the 1st filtration step in the full scale plant Grobbendonk ('BW 1st RSF'), and backwash water from the pilot filter column A1 ('BW A1'). Backwash water was used to obtain a higher bacteria yield. 'BW A1' were sampled when filter media ripening was almost completed (i.e., manganese removal efficiency in the filter was > 90 %).

Measurements and subsequent identification of bacteria, present in 'BW 1st RSF' and 'BW A1' were carried out with "Next generation DNA sequencing", quantitative Polymerase Chain Reaction (qPCR) and Matrix-assisted laser desorption/ionization time-of-flight mass spectrometry (MALDI-TOF MS). Finally, growth tests were performed in a fermenter with selected bacteria from a sample of 'BW A1', to examine if MnO_x could be produced biologically. In addition a pure culture of *Pseudomonas putida* (ATCC 23483, LMG 2321), grown on a selective growth medium (section 5.3.4), was used as a reference.

5.3.1 Next generation DNA sequencing

Samples were taken from both, 'BW 1st RSF' and 'BW A1'. A part of the *16S rRNA* gene (approximately 900 bp) was amplified from these samples, using a eubacterial forward primer GM3 (5'-AGA GTT TGA TCM TGG C-3'), and the universal reverse primer 926r (5'-CCG TCA ATT CMT TTG AGT TT-3') with identifiable sample bar codes. The pyrosequencing analysis of the amplified *16S rRNA* genes was performed using at LGC genomics (Berlin, Germany) with a 454 Life Sciences GS FLX series genome sequencer upgraded to long read length (Roche, The Netherlands). The returned *16S rRNA* gene sequences were analysed, trimmed, aligned, and identified using the metagenomics tool in the software package Bionumercs (Schloss *et al.,* 2011) based on the 454 SOP (Schloss *et al.,* 2011) and the Mothur pipeline software tool (Schloss *et al.,* 2009).

In short, sequences were analysed and sequencing errors were reduced using flowgrams. To reduce computing time a max number of flows were set to 650-900 depending on the number of available sequences (50,000-200,000). Subsequently, sequences were trimmed (arbitrary choices: only sequences with min. lengths of 200 bp and with both primer sequences were selected, tdiffs: 3, and max homop: 8). Sequences were identified against the Silva reference file release 111 (www.arb-silva.de). To display sequence abundance a taxonomic tree was calculated with a minimal percentage of all observation of >1 %.

5.3.2 qPCR

For the qPCR measurements employed to determine *Leptothrix* sp., a Light cycler 480 II of Roche was used. Samples were taken from both, 'BW 1st RSF' and 'BW A1'. 100mL water was filtered through a 0.45 μm membrane filter. The filter was used in the 'powerbiofilm DNA extraction kit' from MoBio (article number: 24000-50). DNA extraction is based on mechanical and chemical lysis. DNA binds to a silica membrane, followed by wash steps. After that, DNA is eluted in 100 μl elution buffer from the MoBio kit. Primers and probe for *Leptothrix* sp. were heterogeneous for its 16S rRNA. Two upstream primers for *Leptothrix* sp. are required to get all relevant species.

Forward primer1 PS-1: 5' ACGGTAGAGGAGCAATC 3'   (Burger *et al.,* 2008a)
Forward primer2 PSP-6: 5' CAGTAGTGGGGGATAGCC 3'   (Burger *et al.,* 2008a)
Reverse primer DSP-6: 5' GCTTTTGTCAGGGAAGAAATC 3'   (Burger *et al.,* 2008a)
Lepto-pr6 forward: 5' CACGCGGCATGGCT 3' *Cy5   (developed by WLN)

The PCR program was as follows: 10 minutes room temperature (uracil glycosylase), 5 minutes 95 °C (denaturation), 50x 30 seconds at 95 °C; 1 minute at 55 °C and 10 seconds at 72 °C. To quantify the number of bacteria, a WLN-III plasmid was developed for target genes. The start concentration of this plasmid is referenced to the mip gene of the standardized Minerva plasmid, and is about 160,000 cDNA/L for *Leptothrix* sp.

To determine the gene copy numbers of *Pseudomonas* spp. and *Pseudomonas putida*, a quantitative PCR (qPCR) protocol, using newly developed primers was used. For the specific detection of *Pseudomonas* spp. primers were developed targeting the 16S rRNA gene, for detection of *P. putida,* the more variable gyrB gene was selected. The primers and probe used for detection of *Pseudomonas* spp. were Pspp16Sf1 (5'-GAG CCT AGG TCG GAT TAG-3'), Psppr3 (5'-CGC TAC ACA GGA AAT TCC AC-3'), and probe PsppP1 (5'-CGC GTG TGT GAA GAA GGT CTT CG-3'). For quantification of *P. putida* the primers PpgyrBf3 (5'-GAC ATC CTG GCC AAG CGT-3') and PpgyrBr3 (5'-CTT CCT GYT CGA TGT AGC-3') and probe PpgyrBp1 (5'-CTG CAR TGG AAY GAC AGC TTC AAC G-3') were selected. Primer specificity and selectivity was analysed and PCR conditions were optimized.

PCR was conducted in 50 μl reaction volumes containing 25 μl of 2x IQ Supermix (Bio-Rad Laboratories BV, The Netherlands), 10 pmol of the forward, reverse primer and probe, 20 mg of bovine serum albumin, and 10 μl of the DNA template. The amplification program consisted of 2 min at 95 °C; 43 cycles of 20 s at 95 °C and 30 sec at 60 °C. Amplification, detection, and data analysis were performed in an iCycler IQ real-time detection system (Bio-Rad Laboratories BV, The Netherlands). The PCR cycle after which the fluorescence signal of the amplified DNA and the probe was detected (threshold cycle [*Cq*]) was used to quantify the gene copy concentration. Quantification was based on comparison of the sample *Cq* value with the *Cq* values of a calibration curve based on known copy numbers of the plasmin containing the 16S rRNA gene of *P. putida* (U70977.1) and the gyrB gene of *P. putida* (HF545867.1).

5.3.3 Matrix-assisted laser desorption/ionization time-of-flight mass spectrometry (MALDI-TOF MS)

With this technique microorganisms can be identified directly after culturing on selective agar media. Spectra were generated with the MALDI-TOF MS biotyper from Brüker Daltonik GmbH, and compared with approximately 4000 spectra in the Brüker Daltonik GmbH database. In a log score from 1 to 3, the MALDI-TOF biotyper defined the similarity of the known and unknown spectra. When the log score is between 2 and 2.3, the genus identification is secure and probable also the species is identified. With a log score > 2.3, it is highly probable that the species is identified. MALDI-TOF MS is based on the chemotaxonomy of microorganisms. This 'fingerprint' is based on identified proteins of the microorganism. These proteins are always present in a living cell and make it possible to characterize microorganisms. In this project we expected *Pseudomonas* to grow in the filter column where Mn^{2+} was oxidized ('BW A1'). Samples, were filtered through a 0.45 μm membrane and incubated on *Pseudomonas* specific agar, a media containing cetrimide and sodium nalidixate to inhibit gram positive bacteria and some gram negatives other than *Pseudomonas*.

Pseudomonas secretes a variety of pigments, including pyocyanin (blue-green), pyoverdine (yellow-green and fluorescent), and pyorubin (red-brown). Coloured colonies on a *Pseudomonas* agar are suspected to be *Pseudomonas,* and they were identified using the MALDI-TOF MS biotyper from Bruker. A single colony of a target organism is put directly on a 96 target plate. After deposition, the spots were overlaid with 1 μl matrix solution (2.5mg α-Cyano-4-hydroxycinnamic solved in 50% acetonitrile, 2.5 % trifluoro acetic acid, 47.5 % ultra-pure water). The matrix opens the cell wall. A laser irradiate the matrix sample, to divide it in little portions of proteins. The matrix evaporate and positive charged proteins become free. In the strong electric field, the positive charged proteins are lined up. So these proteins have the same starting point, before they accelerate in the flight tube to get to their specific time-of-flight corresponding with their specific mass.

5.3.4 Fermenter growth test with selected bacteria to produce biological MnOx

A Bioflow III fermenter from New Brunswick scientific was inoculated with pure cultures of *Pseudomonas putida* (ATCC 23483, LMG 2321), as well as with *P. grimontii* and *P. koreensis*, which were isolated from 'BW A1'. The growth medium used in the fermenter, is described by Jiang *et al.* (2010). Growth and subsequent manganese oxidation in the fermenter was performed for 4 days. Growth of the *Pseudomonas species* was performed by use of the standard *Pseudomonas* agar growth medium (48.4 g agar and 10 ml Glycerol per liter and sterilized for 15 min at 121 °C). The incubation time for growth was 24-48 hrs. at 30 °C. Colonies were identified by MALDI-TOF MS (section 5.3.3) and stored in Pentone-glycerol at - 80 °C. Formation of MnO_x was identified as black deposits and was verified by ICP-MS and SEM-EDX.

5.4 Results and Discussion

5.4.1 Next generation DNA sequencing

Table 5.1 provides an overview of the bacteria population found in backwash water samples from the first stage filter ('BW 1ˢᵗ RSF') and the pilot filter column ('BW A1'). The sequencing results of sample 'BW 1ˢᵗ RSF' are based on 188,241 sequences, whereas the results of sample 'BW A1' are based on 55,298 sequences. The taxonomic trees are shown in Annex A and B.

Table 5.1: Bacteria population in samples of 'BW 1ˢᵗ RSF' and 'BW A1'.

Sample	Identification	% of population
BW 1ˢᵗ RSF	*Gallionella* sp.	97.0
	Other	3.0
BW A1	*Nitrospira* sp.	25,7
	Pseudomonas sp.	14,3
	Gallionella sp.	12,4
	Other	47,6

Table 5.1 shows a clear difference in bacterial population present in sample 'BW 1ˢᵗ RSF' (iron removal filter of the full scale plant) in comparison to bacteria present in 'BW A1' of the freshly ripened (manganese removal) pilot filter column.

The majority (97 %) of bacteria found in 'BW 1ˢᵗ RSF' consisted of *Gallionella* sp. The abundance of *Gallionella* sp. is understandable as iron removal takes place by a biological removal mechanism. Very similar bacteria composition, is present in the feed water to the 2ⁿᵈ stage full scale and thus also in the feed water of the pilot filter column. Identification of bacteria present in 'BW A1', showed that only 12.4 % of the bacteria population was of a *Gallionella species*. *Pseudomonas* sp. and *Nitrospira* sp. represented 14.3 % and 25.7 % of the total population, respectively. Almost half of the population found in 'BW A1' belongs to smaller populations or could not be identified. The presence of the bacterium *Nitrospira* sp. (25.7 %) was expected, because this species is involved in the oxidation of ammonia and specifically conversion of nitrite to nitrate, which takes place in this filter. *Pseudomonas* sp. might be involved in the biological manganese removal process. Literature suggests that besides oxidation of manganese, *Pseudomonas* sp. is able to oxidize ammonia (Daum *et al.,* 1998, Nemergut and Schmidt, 2002). This finding also explains the strong relation between manganese and ammonia oxidation observed in practice (Bruins *et al.,* 2014). The observation that 14,3 % of the bacteria population found in 'BW A1' consists of *Pseudomonas* sp., a potential manganese oxidizing bacterium, supports the assumption that manganese removal starts biologically (Bruins *et al.,* 2015b).

5.4.2 qPCR

Leptothrix sp. and *Pseudomonas* sp., are both able to oxidize dissolved manganese. Their concentration was quantified with qPCR as the number of DNA copies (n [cDNA/L]) present in 'BW A1'. Furthermore the concentration of the species *Pseudomons putida* was quantified with the same technique. Table 5.2 gives an overview of the number of quantified species, expressed as DNA copies/L.

Table 5.2: *Manganese oxidizing bacteria in sample 'BW A1', quantified by qPCR.*

Bacterium	n (cDNA/L)	%
Pseudomonas spp.	2.3×10^{11}	> 99.99 %
Pseudomonas putida	$1.5 * 10^{7}$	< 0.01 %
Leptothrix sp.	3.8×10^{6}	< 0.01 %

From table 5.2 it is clear that from the potential manganese oxidizers the presence of *Pseudomonas* spp. was much more pronounced, than the presence of *Leptothrix* sp. This supports the fact that *Leptothrix* sp. was not found with the Next generation DNA sequencing (section 3.1). Also *Pseudomonas putida* was present in relatively low concentrations, compared to the genus *Pseudomonas*. In literature *Pseudomonas putida* is often associated with biological manganese removal. However, the species *Pseudomonas putida* was not ubiquitous and thus not likely responsible for the fast ripening of manganese removal filters in the pilot testing performed in this study. At the same time it is plausible that other closely related *Pseudomonas* species contributed to this process.

5.4.3 Matrix-assisted laser desorption/ionization time-of-flight mass spectrometry biotyper (MALDI-TOF MS)

Several colonies isolated from sample 'BW A1', were cultured with a *pseudomonas* agar and were identified by using the MALDI-TOF MS biotyper. Table. 5.3 shows an overview of all identified *Pseudomonas* species present in the backwash water of a freshly ripened manganese removal filter.

Table 5.3: *Pseudomonas species identified by MALDI-TOF MS (including Log score).*

Identified species	Log score
Pseudomonas gessardii	2.40
Pseudomonas (libanensis)	2.22
Pseudomonas (synxantha)	2.21
Pseudomonas (veronii)	2.17
Pseudomonas (grimontii) Fig. 5.2	2.15
Pseudomonas (koreensis)	2.14
Pseudomonas (extremorientalis)	2.07
Pseudomonas (marginalis)	2.04
Pseudomonas (tolaasii)	2.03
Pseudomonas (azotoformans)	2.03
Pseudomonas (rhodesiae)	2.00

The genus *Pseudomonas* consists of many very closely related species. No *Pseudomonas putida* was identified in any of the samples with MALDI-TOF MS. This was expected based on the low contribution of the strain *P. putida*, to the total bacterial population determined with qPCR, (< 0.01 %). The list of *Pseudomonas* species (table 5.3) is incomplete, as only a limited amount of colonies are identified. SEM images (Fig. 5.2) show two *Pseudomonas* species, namely *Pseudomonas grimontii and Pseudomonas koreensis* in the backwash water of a freshly ripened manganese removal filter.

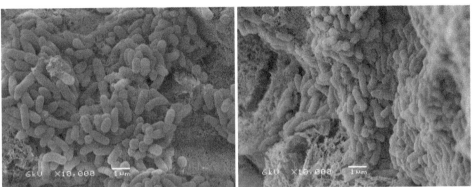

Figure 5.2: *SEM-images (10.000x) of Pseudomonas grimontii (L) and Pseudomonas koreensis (R), isolated from sample 'BW A1'.*

5.4.4 Fermenter growth test with selected bacteria for the biological production of MnO$_x$

Pseudomonas grimontii (log: 2.15) and *Pseudomonas koreensis* (log: 2.14), obtained from 'BW A1', were used as inoculum in a fermenter to investigate their growth and MnO$_x$ production under controlled laboratory conditions. As a reference, a similar growth test was conducted with the laboratory species *Pseudomonas putida* (ATCC 23483, LMG 2321). Although to a very limited extend, *Pseudomonas putida* was able to produce MnO$_x$. Results obtained from the fermenter growth test showed that *Pseudomonas grimontii* and *Pseudomonas koreensis* were not able to oxidize Mn^{2+} producing MnO$_x$, under the performed laboratory conditions.

Summarizing, results from this study show that the population of bacteria present in the backwash water of the 1st stage (iron removal) filter and the freshly ripened (manganese removal) pilot filter column, dramatically changed. Furthermore the presence of *Pseudomonas putida* was very limited (< 0.01 % of the potential manganese oxidizing bacteria present). This indicates that the role of *Pseudomonas putida*, concerning manganese removal at the location Grobbendonk, is limited. However, related *Pseudomonas* species, may play an important role in the process of manganese removal (Table 5.3), taking into account that the Birnessite (MnO$_x$) produced in the pilot filter column during the ripening period was of biological origin (Bruins *et al.*, 2015b). It remains, however, unclear if *Pseudomonas* sp. is the only manganese oxidizing bacterium involved in the initial Mn^{2+} oxidation, or that other species form a microbial consortium, together, responsible for the oxidation of Mn^{2+}. The knowledge that manganese removal in aeration-rapid sand filtration treatment is initiated biologically, together with insight in the manganese oxidizing bacteria species involved, may enable typically long ripening times to be reduced by creating conditions favorable for the growth of these manganese oxidizing species. Therefore, the focus of the follow-up research will be on the inoculation of a consortium of bacteria, identified in manganese removing filters, to enhance filter media ripening. Also conditions supporting the fast growth of Mn^{2+} oxidizing bacteria should be investigated in follow up research.

5.5 Conclusions

From this study it can be concluded that:

- Based on 'next generation DNA sequencing' analyses, the population of bacteria present in backwash water from an iron removal filter (first step filter in a full scale plant), and the freshly ripened pilot manganese removal filter showed a clear population shift from the iron oxidizing species *Gallionella* sp. (97 %) to manganese and nitrite oxidizing species (*Pseudomonas* sp. (14 %) and *Nitrospira* sp. (26 %), respectively). However, it should be noted that 47.6 % of the bacteria population in the manganese removal filter, is still unknown;
- qPCR analysis showed that less than 0.01 % of the genus *Pseudomonas* present, in freshly ripened manganese removal pilot filter column was of the *Pseudomonas putida* species;
- *Pseudomonas* sp. is most likely (one of) the manganese oxidizing bacterium genus that play an important role in the initial stage of the ripening of the manganese removal filters at the full scale GWTP Grobbendonk. However, it is, still unclear whether this bacterium genus is solitary operating or acting as part of a microbial consortium;
- Amongst others, *P. gessardii, P. grimontii and P. koreensis*, closely related *Pseudomonas* species, were detected by the MALDI-TOF MS analysis, and are likely involved in the manganese removal process, possibly as a part of a bacterial consortium;
- Selected *Pseudomonas* species from the ripened filter media column namely *Pseudomonas grimontii* and *Pseudomonas koreensis* were not able to produce MnO_x under controlled laboratory conditions, whereas the reference species *Pseudomonas putida* was able to do so.

5.6 Acknowledgements

This research was financially and technically supported by WLN and the Dutch water companies Groningen (WBG) and Drenthe (WMD). The auteurs are grateful to Mrs. Marsha van der Wiel and Mr. Pim Willemse of WLN for providing qPCR, MaldiTOF MS analysis and performing the growth tests and to Mr. Jelmer Dijkstra of Wetsus for performing SEM-analysis. Thanks also to water company Pidpa (Belgium), for their willingness to share the data from their groundwater treatment plant and to give the opportunity to perform a pilot test at Grobbendonk

5.7 References

Adams, L.F., Ghiorse, W.C., 1985. Influence of Manganese on Growth of a Sheathless Strain of *Leptothrix discophora*. Aplied and environmental Microbiology, 49.3, 556 –562.

Barger J.R., Tebo, B.M., Bergmann, U., Webb, S.M., Glatzel, P., Chiu, V.Q., Villalobos, M., 2005. Biotic and abiotic products of Mn (II) oxidation by spores of the marine *Bacillus* sp. strain SG-1. American Mineralogist, 90, 143-154.

Barger, J.R., Fuller. C.C., Marcu. M.A., Brearly A., Perez De la Rosa M., Webb S.M., Caldwell W.A., 2009. Structural characterization of terrestrial microbial Mn oxides from Pinal Ckeek, AZ. Ceochimica et Cosmochimica Acta 73, 889-910.

Boogerd, F.C., De Vrind J.P.M., 1987. Manganese oxidation by *Leptothrix discophora*. Journal of bacteriology, 169.2, 489-494.

Brouwers, G.J., Vrind de J.P.M, Corstjens P.L.A.M., Cornelis P, Baysse C, Vrind de Jong E.W., 1999. CumA, aGene Encoding a multicopper oxidase, is involved in Mn^{2+}oxidation in *Pseudomonas putida* GB-1. Applied and environmental microbiology , 65:4, 1762-1768.

Brouwers G.J., Vijgenboom E., Corstjens P.L.A.M., de Vrind J.P.M., de Vrind-de Jong E.W., 2000. Bacterial Mn^{2+} oxidizing systems and multicopper oxidases: an overview of Mechanisms and Functions. Geomicrobiology Journal, 17.1, 1-24.

Bruins, J.H., Vries, D., Petrusevski, B., Slokar, Y.M., Kennedy, M.D., 2014. Assessment of manganese removal from over 100 groundwater treatment plants. Journal of Water Supply: Research and Technology-AQUA, 63.4, 268-280.

Bruins, J.H., Petrusevski, B., Slokar, Y.M., Kruithof, J.C., Kennedy, M.D., 2015a. on line. Manganese removal from groundwater: characterization of filter media coating. Desalination & Water Treatment, in press, doi: 10.1080/19433994.2014.927802.

Bruins, J.H., Petrusevski, B., Slokar Y.M., Huysman, K., Joris, K., Kriuithof J.C., Kennedy, M.D., 2015b. Biological and physicochemical formation of Birnessite during the ripening of manganese removal filters. Water research 69, 154-161.

Burger, M.S., Krentz, C.A., Mercer, S.S., Gagnon, G.A., 2008a. Manganese removal and occurrence of manganese oxidizing bacteria in full-scale biofilters. Journal of Water Supply: Research and technology-AQUA 57.5, 351-359.

Burger, M.S., Mercer, S.S., Shupe, G.D., Gagnon, G.A., 2008b. Manganese removal during bench-scale biofiltration. Water Research 42, 4733-4742.

Caspi, R., Tebo, B.M., Haygood, M.G., 1998. C-Type Cytochromes and Manganese Oxidation in *Pseudomonas putida* MnB1. Applied and environmental Microbiology, 64.10, 3549-3555.

Corstjens, P.L.A.M., de Vrind, J.P.M., Goosen, T., de Vrind-de Jong, E.W., 1997. Identification and molecular analysis of the Leptothrix discophora SS-1 mofA gene, a gene putatively encoding a manganese-oxidizing protein with copper domains. Geomicrobiology Journal, 14.2, 91-108.

DePalma, S.R., 1993. Manganese oxidation by *Pseudomonas putida*, PhD-thesis, Harvard University, Cambridge, USA.

De Schamphelaire, L. , Rabaey, K. , Boon, N. , Verstraete, W. and Boeckx, P., 2007. Minireview: The potential of enhanced manganese redox cycling for sediment oxidation. Geomicrobiology Journal, 24: 7, 547-558.

Daum, M., Zimmer, W., Papen, H., Kloos, K., Nawrath, K., Bothe, H., 1998. Physiological and Molecular Biological Characterization of Ammonia Oxidation of the Heterotrophic Nitrifier *Pseudomonas putida*. Current Microbiology, 37, 281–288.

El Gheriany I.A., Bocioaga B., Anthony Hay A.D., Ghiorse W.C., Shuler M.L., Lion L.W., 2009. Iron Requirement for Mn (II) Oxidation by *Leptothrix discophora* SS-1. Applied and Environmental Microbiology, 75.5, 1229-1235.

Farnsworth, C.E., Voegelin, A. and Hering, J.G., 2012. Manganese oxidation induced by water table fluctuations in a sand column. Environ. Sci. Technol. 46, 277-284.

Feng, A., Tan, W., Liu, F. Huang, Q., and Liu, X., 2005. Pathways of Birnessite formation in alkali medium. Sci. China Ser. D., 48.9, 1438-1451.

Fleming, E.J., Langdon, A.E., Martinez-Garcia, M., Stepanauskas, N.J., Poulton, Masland, E.D.P., Emerson, D., 2011. What's New Is Old: Resolving the Identity of Leptothrix ochracea Using Single Cell Genomics, Pyrosequencing and FISH. PLoS ONE, 6.3, 1-10.

Forrez, I., Carballa, M., Verbeken, K., Vanhaecke, L., Schlüsener, M., Ternes, T., Boon, N., Verstrete, W., 2010. Diclofenac oxidation by biogemic manganese oxides. Environmental science & technology, 44.9, 3449-3454.

Geszvain, K., McCarthy, J.K., Tebo, B.M., 2013. Elimination of Manganese (II, III) Oxidation in *Pseudomonas putida* GB-1 by a Double Knockout of Two Putative Multicopper Oxidase Genes. Applied and Environmental Microbiology, 79 .1, 357–366.

Gounot A-M., 1994. Microbial oxidation and reduction of manganese: Consequences in groundwater and applications. FEMS Microbiology reviews 14, 339-350.

Hope C.K., Bott T.R., 2004. Laboratory modelling of manganese biofiltration using biofilms of *Leptothrix doscophora*. Water research 38, 1853-1861.

Jiang, S., Kim, D-G., Kim, J., Ko, S-O., 2010. Characterization of the biogenic manganese oxides produced by *Pseudomona putida* strain MnB1. Environ. Eng. Res., 15.4, 183-190.

Kim, S.S., Bargar, J.R., Nealson, K.H., Flood, B.E., Kirschvink, J.L., Raub, T.D., Tebo, B.M., Villalobos, M., 2011. Searching for Biosignatures Using Electron Paramagnetic Resonance (EPR) Analysis of Manganese Oxides. Astrobiology, 11.8, 775-786.

Mann, S., Sparks. N.H.C., Scoot G.H.E., Vrind- De Jong E.W., 1988. Oxidation of Manganese and formation of Mn_3O_4 (Hausmannite) by spore Coats of a marine Bacillus sp. Applied and environmental microbiology, 54.8, 2140-2143.

Meerburg, F., Hennebel, T., Vanhaecke, L., Verstraete, W., Boon, N., 2012. Diclofenac and 2-anilinophenylacetate degradation by combined activity of biogenic manganese oxides in silver. Microbial technology, 5.3, 388-398.

Nemergut, D.R., Schmidt, S.K., 2002. Disruption of *narH*, *narJ* and *moaE* inhibits heterotrophic nitrification in *Pseudomonas* strain M19. Applied and environmental biology, 68.12, 6462-6465.

Post, J.E. and Veblen, D.R., 1990. Crystal structure determinations of synthetic sodium, magnesium, and potassium birnessite using TEM and the Rietveld method. American mineralogist, 75, 477-489.

Post, J. E., 1999. Manganese oxide minerals: Crystal structures and economic and environmental significance. Proceedings of the National Academy of Sciences USA, 96, 3447-3454.

Sabirova, J.S., Cloetens, L.F.F., Vanhaecke, L., Forrez, I., Verstraete, W., Boon, N., 2008. Manganese-oxidizing bacteria mediate the degradation of 17α-ethinylestradiol. Microbial Biotechnology, 1.6, 507-512.

Schloss, P.D. Westcott, S.L., Ryabin, T., Hall, J.R., Hartmann, M., Hollister, E.B., Lesniewski, R.A., Oakly, B.B., Parks, D.H., Robinson, C.J., Sahl, J.W., Stres, B., Thallinger, G.G., Van Horn, D.J., Weber, C.J., 2009. Introducing Mothur: Open source, platform-independent, community-supported software for describing and comparing microbial communities. Appl. Environ Microbiol, 75 (23), 7537-7541.

Schloss, P.D., Gevers, D., Westcott, S.L., 2011. Reducing the effects of PCR amplification and sequencing artifacts on 16S rRNA-based studies. PloS One. 6 (12): e27310, doi:10.1371/journal.pone.0027310.

Stumm, W. Morgan, J.J., 1996. Aquatic Chemistry, chemical equilibria and rates, 3rd ed. Wiley New York.

Tebo, B.M., Marger J.R., Clement B.G., Dick G.J., Murray K.J., Parker. D., Verity R., Webb S.M., 2004. Biogenic Manganese oxides: Properties and mechanisms of formation. Annu. Rev. Earth Planet Sci., 32, 287-328.

Tebo, B.M., Johnson H.A., McCarthy J.K., Templeton A.S., 2005. Geomicrobiology of manganese (II) Oxidation. TRENDS in Microbiology, 13.9, 421-428.

Tekerlekopoulou, A.G., Vasiliadou I.A., Vayenas D.V., 2008. Biological manganese removal from potable water using trickling filters. Biochemical Engineering Journal 38, 292-301.

Vandenabeele, J., de Beer, D., R. Germonpré, R., Verstreate, W., 1992. Manganese oxidation by microbial consortia from sand filters. Microbial Ecology 24, 91-108.

Vandenabeele J., 1993. Manganese removal by microbial consortia from rapid sand filters treating water containing Mn^{2+} and NH_4^+, PhD thesis, Ghent, Belgium.

Verstraete, W., 2013, Personal communication, Laboratory of microbial ecology and technology, department of biochemical and microbial technology, Ghent University, Belgium.

Villalobos, M., Toner, B., Bargar, J., Sposito, G., 2003. Characterization of the manganese oxide produced by *Pseudomonas putida* strain MnB1. Geochimica et Cosmochimica Acta, 67.14, 2649-2662.

Villalobos, M., Lanson, B., Manceau, A., Toner, B., Sposito, G., 2006. Structural model for the biogenic Mn oxide produced by *Pseudomonas putida*. American Mineralogist, 91, 489-502.

Vrind de, J.P.M., Boogerd, F.C., Vrind de – Jong de, E.W., 1986. Manganese reduction by a marine Bacillus Species. Journal of Bacteriology, 167.1, 30-34.

Webb, S.M., Tebo, B.M., Bargar, J.R., 2005a. Structural characterization of biogenic Mn oxides produced in seawater by marine *bacillus* sp. strain SG-1. American mineralogist, 90, 1342-1357.

Webb, S.M., Dick, G.J., Bargar, J.R., Tebo, B.M., 2005b. Evidence for the presence of Mn (III) intermediates in the bacterial oxidation of Mn (II). Proceedings of national academic science, 102.15, 5558-5563.

Annex A
Next generation DNA sequencing, taxonomic trees of sample 'BW 1st RSF' (based on 188,241 sequences)

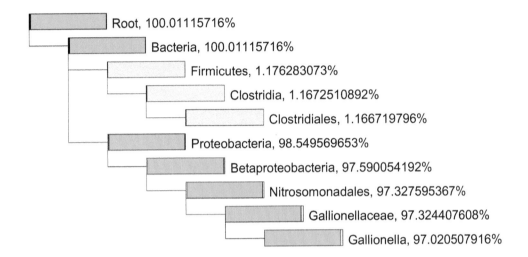

Root, 100.01115716%

Bacteria, 100.01115716%

Firmicutes, 1.176283073%

Clostridia, 1.1672510892%

Clostridiales, 1.166719796%

Proteobacteria, 98.549569653%

Betaproteobacteria, 97.590054192%

Nitrosomonadales, 97.327595367%

Gallionellaceae, 97.324407608%

Gallionella, 97.020507916%

Annex B
Next generation DNA sequencing, taxonomic trees of sample 'BW A1' (based on 55,298 sequences)

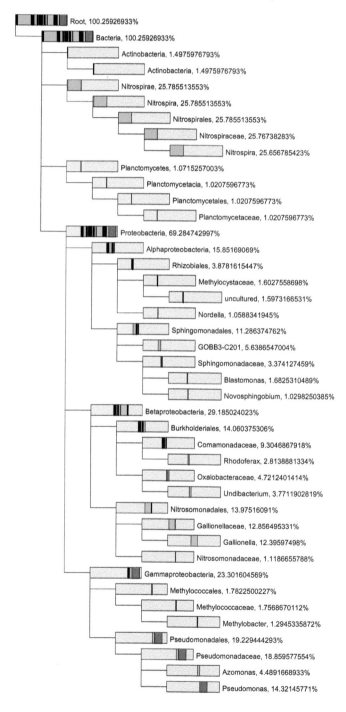

Figure: *(top) GWTP Grobbendonk (2015, Koen Joris, Pidpa) and (bottom) GWTP De Punt (2014, WBG)*

6 REDUCTION OF RIPENING TIME OF FULL SCALE MANGANESE REMOVAL FILTERS WITH MANGANESE OXIDE COATED MEDIA

Main part of this chapter was published as:

Jantinus H. Bruins, Branislav Petrusevski, Yness M. Slokar, Koen Huysman, Koen Joris, Joop C. Kruithof, Maria D. Kennedy (2015). Reduction of ripening time of full scale manganese removal filters with manganese oxide coated media. Journal of Water Supply: Research and Technology – AQUA 64.4: 434 – 441

6.1 Abstract

Effective manganese removal by conventional aeration-filtration with virgin filter media requires a long ripening time. The aim of this study was to assess the potential of manganese oxide-coated media to shorten the ripening time of filters with virgin media, under practical conditions. A full scale filter filled with virgin sand and a full scale filter filled with anthracite/sand were operated at two groundwater treatment plants, in parallel with (full scale) test filters, with an additional layer of Manganese Oxide-Coated Sand (MOCS) or Manganese Oxide-Coated Anthracite (MOCA). Significantly different ripening times were observed to achieve an effective manganese removal: 55 days for a filter with virgin sand and 16 days for a filter with virgin anthracite/sand, respectively. The observed differences could be attributed to different feed water quality, different iron loading, and backwashing intensity and frequency. In batch experiments fresh MOCA and MOCS showed good manganese adsorptive properties. Addition of a shallow layer of fresh MOCA in test filters eliminated the ripening time, while a layer of aged MOCS did not significantly shorten the ripening period. The poor performance of the aged MOCS was likely caused by changed properties of aged and dried MOCS that had lost its adsorption capacity, the auto-catalytic activity and the biological activity.

Keywords: filter media, filter ripening, groundwater treatment, manganese oxide coating, manganese removal

6.2 Introduction

In some countries (*e.g.*, the US, and Central and Eastern Europe) an efficient manganese removal is commonly achieved by pre-oxidation with strong oxidants, such as O_3, Cl_2, ClO_2, $KMnO_4$, followed by rapid sand filtration. Use of strong oxidants for manganese removal is not desirable due to the potential formation of harmful oxidation by-products, as well as costs and risks associated with the usage and handling of chemicals. In some cases pre-oxidation is combined with filtration through a filter bed with a manganese adsorbent, most frequently manganese green sand. This treatment can be very effective but it requires continuous or intermittent regeneration typically with potassium permanganate (Knocke *et al.*, 1991). It is in view of the above mentioned disadvantages, that the removal of manganese from groundwater in The Netherlands and Belgium is normally achieved with conventional aeration-filtration treatment, also called contact filtration. Under common groundwater conditions (*e.g.*, low pH), manganese removal may be initiated by bacterial activity during aeration-filtration (Diem & Stumm, 1984; Burger *et al.*, 2008). Although aeration-filtration is efficient and cost effective, it is in practice, however, frequently associated with a major drawback:

- very long ripening times of virgin filter media; several weeks to more than a year (Fig. 6.1), are required to achieve an efficient manganese removal (Buamah *et. al.,* 2009a; Cools, 2010; Krull, 2010).

Ripening of filter media (for manganese removal), is defined as the development of properties to auto-catalytically adsorb and subsequently oxidise Mn^{2+}, without the use of strong oxidants, such as Cl_2, O_3 and $KMnO_4$.

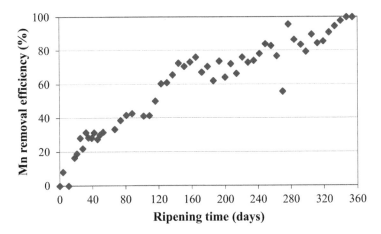

Figure 6.1: *An example of a very long ripening time of virgin filter media for manganese removal in a Dutch full scale groundwater treatment plant (GWTP Baanhoek, Evides Water Company).*

From Fig. 6.1, it can be seen that efficient manganese removal in a filter with virgin sand was not achieved until after almost one year of continuous filter operation. However, such a long ripening time is exceptional, typically it takes 1 to 4 months to achieve an efficient manganese removal.

Many authors (Hu *et al.,* 2004a; Kim & Jung, 2008; Kim *et al.,* 2009) describe the potential of Manganese Oxide-Coated Sand (MOCS) to adsorb dissolved manganese from (ground)water. It was also reported that removal of Mn^{2+} in filters with anthracite is enhanced by development of 'catalytic oxide layers' on aged anthracite, due to formation of Manganese Oxide-Coated Anthracite (MOCA) (Sahabi *et al.,* 2009). Buamah *et al.* (2008) suggested that the performance of conventional manganese removal plants could be improved by introducing manganese and/or iron (hydro)oxide-rich filter media into rapid sand filters.

The primary aim of the study reported in this paper was to examine if the long ripening time typically required to achieve an effective manganese removal with virgin filter media, in full scale conventional aeration-filtration treatment plants, could be substantially reduced by addition of a MOCS or MOCA layer.

6.3 Materials and Methods

6.3.1 MOCS and MOCA

The manganese oxide-coated filter media used in this research were obtained from two full scale groundwater treatment plants (GWTPs). MOCS was obtained from GWTP De Punt (Water Supply Company Groningen, The Netherlands), and MOCA was obtained from GWTP Grobbendonk (Pidpa water supply company, Belgium). It was shown that both MOCS and MOCA coatings contain a Birnessite type of manganese oxide (Bruins *et al.,* 2014a).

For all batch adsorption experiments and full scale filter experiments fresh MOCA or MOCS were taken directly from an operating, ripened full scale manganese removal filter. However, MOCS was stored in the open air for several months prior to the full scale filter experiments.

6.3.2 Physical and chemical properties of MOCS and MOCA

Chemical composition of the MOCS and MOCA coating was determined by boiling the media in 3 M HNO_3, followed by analysis for Fe, Mn, Ca, Si and Al with inductively coupled plasma mass spectrometry (ICP-MS) according to NEN-EN-ISO 17294-2 (NEN 2004).

The pH of point of zero charge (pH_{PZC}), *i.e.*, the surface charge of coated filter media depending on structural deficits, unbalanced bonds and the presence of protons, was determined by a mass titration method (Fiol & Vilaescusa, 2009).

6.3.3 Batch adsorption experiments

To determine the MOCS and MOCA manganese adsorption capacity, batch adsorption isotherm experiments were carried out. Model water used in these experiments contained 1 mmol/L HCO_3^- and 2 mg/L Mn^{2+} in demineralised water; the pH was adjusted to 7 with 0.1 M HCl. The bottles containing model water and five different concentrations of either MOCS or MOCA in granular form were agitated on an Innova 2100 shaker at 100 rpm for 48 h. Prior to measuring the final concentration of manganese, the samples were filtered through a 0.45 µm membrane filter, and acidified to preserve them for the analyses. The manganese concentration was measured, and results were plotted as a Freundlich adsorption isotherm.

The results obtained from the batch adsorption experiments of the two manganese-coated filter media were compared with those obtained with a commercial manganese adsorbent Aquamandix (Aqua-techniek, The Netherlands).

6.3.4 Full scale filter runs

Full scale filter runs were conducted at GWTPs where the manganese-coated filter media were obtained. In total, six different combinations of filter media were used (Fig. 6.2). Two filters were operated at GWTP De Punt (Fig. 6.2[A-B]), and the other four (Fig. 6.2[C-F]) at GWTP Grobbendonk.

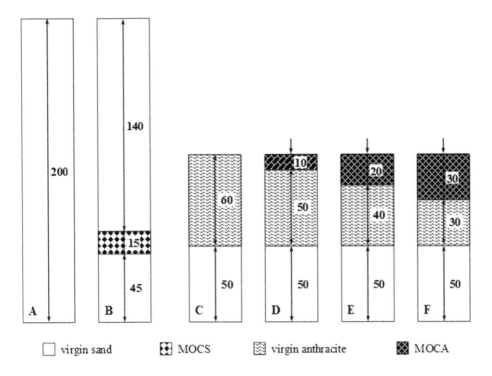

virgin sand MOCS virgin anthracite MOCA

Figure 6.2: *Schematic presentation of filter media layers in the six full scale filters included in the study (all values in cm).*

Of the two filters at GWTP De Punt, one filter (Fig. 6.2A) was filled with virgin quartz sand, commonly applied at this plant. This filter served as a reference filter. The second filter (Fig. 6.2B) was filled with the same virgin sand, however, a 15 cm layer of sand was replaced by MOCS. The MOCS was aged and dried prior to use in the full scale filter run. Based on practical experience, the MOCS layer was placed at the level where manganese removal in ripened filters is observed at this facility.

The effect of the MOCA layer on the ripening of virgin filter media was studied at GWTP Grobbendonk. This plant utilises filters with dual media - anthracite and sand - which is how the reference filter was prepared (Fig. 6.2C). In the other three (full scale) test filters (Fig. 6.2D-F), part of the top anthracite layer was replaced by 10, 20 and 30 cm MOCA respectively. MOCA was placed on the top of the anthracite layer, because at this level manganese removal is observed at this water treatment plant.

The composition of the feed water for the full scale experiments at the two GWTPs is given in Table 6.1. At GWTP De Punt, groundwater was aerated prior to the test filters. At GWTP Grobbendonk the treatment consists of a 1st stage rapid sand filtration (aeration and biological adsorptive iron removal), a pH correction with milk of lime and 2nd stage dual media filtration. The feed water for the experiments at GWTP Grobbendonk was the water after pH correction.

Table 6.1: *Composition of feed water during the experiments with full scale filters.*

Parameter	Unit	De Punt	Grobbendonk
Iron	mg/L	4.5 - 6.9	0.03 - 0.14
Manganese	mg/L	0.18 - 0.25	0.12 - 0.18
Ammonium	mg/L	0.29 - 0.78	< 0.05 - 0.23
pH	[-]	7.3 -7.5	7.5 - 7.6
Oxygen	mg/L	8 - 10	> 10
Redox potential	mV	-50 to +50	+ 200 to +300

From Table 6.1 it is evident that the feed water quality at the two test locations differed significantly. In particular the difference in water quality parameters that are known to influence manganese removal (Fe^{2+}, NH_4^+ concentrations, pH and redox potential) should be noted (Bruins *et al.*, 2014b).

Table 6.2: *Process design parameters and operational conditions of full scale manganese removal experiments at GWTP De Punt and GWTP Grobbendonk.*

Parameter / condition	Unit	De Punt	Grobbendonk
Type of aeration	-	spray	cascade
Position of filter	-	'pre-filter'	'post-filter'
Type of filtration	-	down flow	down flow
Type of filter media	-	quartz sand	anthracite / quartz sand
Grain size fraction virgin media	mm	1.8 - 2.4	0.8-1.8 / 0.4-0.8
Filter area	m²	12.5	37.5
Filter bed height	m	2	1.1 (0.6+0.5)
Flow per filter	m³/h	60	190
Filtration rate	m³/m².h	4.8	5.0
Empty bed contact time	min	25	13.2
Backwash (BW) criterion	-	head loss	head loss
Backwash frequency	n/week	2	0.5
Filter bed expansion during BW	-	no	yes (anthracite)
Filtered volume between BW	m³ per filter run	5,000 – 7,000	10,000
Iron loading per filter run (FR)	kg Fe/m².FR	2.5	< 0.1

Table 6.2 depicts an overview of process design parameters and operational conditions applied during the test filter runs at both locations. From this table it is evident that especially 'Iron loading per filter run' is substantially different (De Punt: 2.5 kg Fe/m².FR and Grobbendonk: < 0.1 kg Fe/m².FR). Iron loading is known to influence manganese removal (Bruins *et al.*, 2014b). In Table 6.3 backwash procedures at both GWTPs are listed.

Table 6.3: *Backwash procedures applied at full scale plants De Punt and Grobbendonk.*

GWTP De Punt			GWTP Grobbendonk		
V_f (m³/m².h)		Duration	V_f (m³/m².h)		Duration
Water	Air	Minutes	Water	Air	Minutes
-	24	1	10.6	-	0.33
-	52	5	-	60	1
13.6	28	3	23.8	-	5
24	28	5	-	-	-
24	-	8	-	-	-
13.6	-	3	-	-	-

6.4 Results and discussion

6.4.1 Ripening of virgin filter media in reference filters

In Fig. 6.3 the ripening times of the two full scale reference filters filled with virgin sand (GWTP De Punt) and anthracite/sand (GWTP Grobbendonk) are shown.

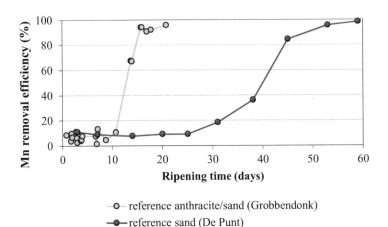

—○— reference anthracite/sand (Grobbendonk)
—●— reference sand (De Punt)

Figure 6.3: *Comparison of the Mn removal efficiency (%) as a function of the filter ripening time for the two reference full scale filters: (virgin) anthracite/sand, at GWTP Grobbendonk (pH: 7.5 to7.6; redox potential: +200 to +300 mV; filtration rate: 5.0 m3/m2.h) and (virgin) sand, at GWTP De Punt (pH: 7.3 to 7.5; redox potential: -50 to +50 mV; filtration rate: 4.8 m3/m2.h).*

As seen from Figure 6.3, the ripening time required to reach > 90% manganese removal of the single media reference filter at GWTP De Punt was about 3.5 times longer (55 days) than that of the dual media reference filter at GWTP Grobbendonk (16 days). The observed difference was attributed to both the different feed water quality and the difference in applied operational conditions (*e.g.*, backwashing pattern, intensity and frequency).

6.4.2 Effect of water quality parameters

It has been reported that ferrous iron competes with Mn^{2+} for adsorption sites on filter media (Hu *et al.,* 2004a, b). Feed water at GWTP De Punt had a 50 to 150 times higher Fe^{2+} concentration than the GWTP Grobbendonk feed (Table 6.1). At GWTP De Punt (in combination with slightly lower feed water pH), the high Fe^{2+} concentration caused a more pronounced competition for available adsorption sites by Mn^{2+}. It has been reported that the presence of iron hydro-oxide layers in the filter media coating could support the Mn^{2+} adsorption (Buamah, 2009). However iron hydro-oxide has a much lower Mn^{2+} adsorption capacity than manganese (hydro-) oxides (Buamah *et al.,* 2008). Therefore formation of iron hydro-oxide layers in the filter bed zone where manganese is removed should be prevented in practice.

The feed water concentration of Fe^{2+} also determines an important operational condition for manganese removal, *i.e.,* iron loading per filter run. Feed water at GWTP De Punt contained a significantly higher Fe^{2+} concentration, resulting in much higher iron loading per filter run than at GWTP Grobbendonk (2.5 and < 0.1 kg Fe/m^2, respectively). In addition, this much higher iron loading at GWTP De Punt required an approximately 4 times more frequent backwashing than at GWTP Grobbendonk. Backwashing results in a partial removal of MnO_x from the coating, while the presence of MnO_x is essential for an effective manganese removal. Intensive backwashing can also cause a substantial removal of the biological activity (*e.g.,* by removal of bacteria from filter media), which may play an important role in the process of manganese adsorption and oxidation (Vandenabeele *et al.,* 1992; Katsoyiannis & Zouboulis, 2004; Tebo *et al.,* 2004). Partial loss of MnO_x and biological activity is even more pronounced for backwashing with combined water and air flushing. Both the frequency and intensity of the backwashing at GWTP De Punt were more detrimental for the filter media, resulting in a longer ripening time of the filters at this location.

Another water quality parameter playing an important role in manganese removal is the pH. To achieve an effective manganese removal the pH should preferably be above 7.1 (Bruins *et al.,* 2014b). In general, the higher the pH, the better the manganese removal. From the pH values of feed water at both GWTPs it can be concluded that Grobbendonk water (pH 7.5 - 7.6) provided slightly better conditions for an effective manganese removal than GWTP De Punt water (pH 7.3 - 7.5).

NH_4^+ removal efficiency is another parameter that shows a strong positive correlation with manganese removal (Bruins *et al.,* 2014b). The presence of NO_2^- (due to incomplete NH_4^+ removal) may not only prevent effective Mn^{2+} removal, but may even cause manganese leaching, by reducing already adsorbed and oxidised MnO_x back to Mn^{2+} (Vandenabeele *et al.,* 1995). The NH_4^+ concentration in the filtrate of the two filters at GWTP De Punt was substantially higher than at GWTP Grobbendonk especially during the filter start-up, which could explain the much longer ripening time of filters at the De Punt location.

Finally, oxidation of (adsorbed) manganese takes place more easily at a higher redox potential (Stumm & Morgan, 1996; Scherer & Wichmann, 2000; Flemming *et al.,* 2004). The redox potential of the feed water at GWTP De Punt was much lower compared to that of the feed water at GWTP Grobbendonk (- 50 / + 50 mV and +200 / +300 mV, respectively), partly caused by the presence of NH_4^+ in the filtrate, which was much higher at the De Punt location. As a consequence, even if Mn^{2+} was adsorbed it was not as effectively oxidised and therefore possibly desorbed, resulting in a longer ripening time at the De Punt location.

Although the feed water quality at both locations is suitable to achieve effective manganese removal in a conventional aeration-filtration system once the filters are ripened, the conditions to achieve shorter filter ripening times were found to be more favourable at the Grobbendonk location.

In summary, much faster ripening of virgin filter media with respect to complete manganese removal at GWTP Grobbendonk can be attributed to the combined effect of the following parameters crucial for manganese removal:

- more favourable feed water quality (lower Fe^{2+} and NH_4^+ concentrations, higher pH and redox potential).
- more favourable operational conditions (lower iron loading per filter run, lower backwash frequency and intensity).

6.4.3 MOCS and MOCA characterisation and batch adsorption experiments

Table 6.4 depicts the physical characteristics of MOCS, MOCA, and Aquamandix (AQM), and coating composition of MOCS and MOCA.

Table 6.4: *Physical characteristics of MOCS, MOCA and AQM, and the coating composition of MOCS and MOCA.*

	Parameter	Unit	MOCS	MOCA	AQM*
Physical properties	Bulk density	kg/L	1.177	0.650	2.000
	Particle density	kg/L	2.326	1.176	3.600
	Porosity	%	49.4	44.7	44.4
	Grain size $(d_{10}-d_{90})$	mm	1.6 - 3.1	0.8 - 1.5	1.0 - 2.0
	Uniformity coefficient	[-]	1.58	1.52	-
	pH_{PZC}	[-]	7.2	8.0	5.0
Coating composition	Mn	mg/g	12.8	13.5	-
	Fe	mg/g	158	2.22	-
	Ca	mg/g	8.85	2.45	-
	Si	mg/g	14.5	0.28	-
	Al	mg/g	0.47	0.25	-

* According to the supplier, Aquamandix consists of 78% $MnO2$, 6.2% $Fe2O3$, 5.2% $SiO2$, 3.1% $Al2O3$.

From Table 6.4 it can be seen that the grain size of MOCA is considerably smaller than that of MOCS. As a consequence, MOCA has a larger geometric surface area, enhancing the adsorption capacity. Another characteristic indicating a better adsorption capacity of MOCA is the coating composition. The most pronounced difference in chemical composition of the two media is the iron content. Iron is present as iron (hydr)oxide, whereas manganese is present as manganese oxide (MnO_x). Although both oxides can absorb Mn^{2+}, iron (hydr)oxide has a significantly lower manganese adsorption capacity (Buamah *et al.*, 2008). Therefore it was expected that MOCA with a more than 70 times lower iron content, would adsorb Mn^{2+} better. On the other hand the pH_{PZC} of MOCS was significantly lower compared to MOCA. This suggests that MOCS will have better adsorptive properties over a wider pH range for positively charged ions such as Mn^{2+}.

In Table 6.5 the Freundlich adsorption isotherm constants for manganese adsorption on MOCS, MOCA, and Aquamandix are given.

Table 6.5: *Freundlich adsorption isotherm constants for Mn²⁺ adsorption on MOCS, MOCA and Aquamandix.*

Constant	Adsorbent		
	MOCS	MOCA	Aquamandix
K [(mg/g) / (mg/L)]	0.45	0.91	0.90
$1/n$	1.31	1.34	1.38
r^2	0.91	0.91	0.96
q_e (mg/g) at C_e=0.2 mg/L	0.132	0.276	0.280
q_e (g /L) at C_e=0.2 mg/L	0.155	0.179	0.560

From Table 6.5 it can be seen that manganese adsorption capacities, q_e expressed per unit weight of adsorbent, for MOCS and MOCA are very different (0.132 and 0.276 mg/g ads., respectively, at C_e of 0.2 mg/L Mn²⁺). When expressed per unit volume, however, adsorptive capacities of MOCS and MOCA were found to be similar (0.155 and 0.179 mg Mn²⁺/L of MOCS and MOCA, respectively, at C_e of 0.2 mg Mn²⁺/L). A much higher adsorption capacity per volume of adsorbent was found for Aquamandix. However this commercial adsorbent has no auto-catalytic oxidation properties, thus once the adsorption capacity is exhausted, manganese removal stops (Buamah, 2009).

Based on the adsorption capacities, q_e reported in Table 6.5, the (calculated) theoretical manganese adsorption capacities of the MOCS and MOCA layers, placed in the full scale test filters are calculated Table 6.6.

Table 6.6: *Calculated theoretical adsorptive capacity of MOCS and MOCA layers with associated duration of filter run (FR) time before the manganese breakthrough.*

Filter	Filter bed media	Capacity[1] (kg)	FR time before Mn breakthrough[2] (hrs)
A	reference sand	0	-
B	sand + 15 cm MOCS	0.29	24 - 48
C	reference anthracite/sand	0	-
D	anthracite/sand + 10 cm MOCA	0.67	24 - 48
E	anthracite/sand + 20 cm MOCA	1.34	48 - 96
F	anthracite/sand + 30 cm MOCA	2.01	72 - 144

1 At Ce=0.2 mg/L
2 For Vf between 2.5 - 5 m³/m².h

This Table (6.6) also shows the expected operational times of the filters before manganese breakthrough, assuming that adsorption was the only manganese removal mechanism ignoring the catalytic effect associated with adsorption and oxidation of adsorbed manganese. For a 15 cm MOCS layer (De Punt), the Mn²⁺ adsorption capacity is approximately 0.29 kg, with an expected breakthrough after 24 to 48 hrs, whereas these values for a 10 cm layer of MOCA (Grobbendonk) are respectively 0.67 kg Mn²⁺ and breakthrough also after 24 to 48 hrs.

6.4.4 Ripening of full scale filters with the addition of MOCS and MOCA layers

Manganese removal during the ripening time of the two full scale filters at GWTP De Punt are shown in Fig. 6.4(L). Based on batch adsorption experiments conducted with fresh MOCS (Tables 6.5 and 6.6), it was expected that the MOCS layer in the full scale filter would effectively remove Mn^{2+} at least during the first 24 - 48 hrs. However, the addition of 15 cm of MOCS did not have a significant impact on the ripening time of the filter with respect to the manganese removal. The reason for the poor Mn^{2+} removal is probably attributed to the difference in adsorptive properties of MOCS used in the batch adsorption experiments (freshly taken from a running ripened filter), and in the filter runs (dried and stored for several months before use). The layered structure of Birnessite may have irreversibly collapsed, decreasing the number of available adsorptive sites (Post, 1999). Furthermore, long exposure to air may have resulted in (complete) oxidation of the auto-catalytically active Birnessite into not auto-catalytically active Pyrolusite (MnO_2). Storage of the MOCS could also have resulted in a loss of biological activity (*e.g.*, by dying of bacteria present on MOCS), which may play an important role in initiating oxidation of manganese adsorbed on filter media (Vandenabeele *et al.*, 1992; Katsoyiannis & Zouboulis, 2004; Tebo *et al.*, 2004). Drying and storage of the MOCS used in the full scale filter most likely caused loss of a substantial part of its original adsorption capacity. Besides, it is most likely the stored MOCS had lost its auto-catalytic and biological activity.

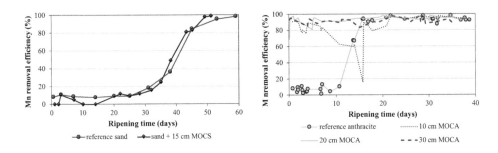

Figure 6.4: *Mn removal efficiency (%) as a function of the ripening time of two (full scale) test filters at GWTP De Punt (L) - (pH: 7.3 to 7.5; redox potential: -50 to +50 mV; filtration rate: 4.8 m³/m².h) and four full scale test filters at GWTP Grobbendonk (R) - (pH: 7.5 to7.6; redox potential: +200 to +300 mV; filtration rate: 5.0 m³/m².h).*

In Fig. 6.4(R) the results of the experiments with four full scale test filters at GWTP Grobbendonk are shown. At this location the filter ripening time of the reference anthracite/sand filter was compared to the ripening times of three test filters containing MOCA layers of different thicknesses (Figs. 6.2D-F).

Results obtained with the test filters containing a layer of MOCA showed a very high (> 90%) manganese removal from the start of the filter run, irrespective of the thickness of the MOCA layer. As a comparison, the manganese removal efficiency in the reference filter without a MOCA layer was approximately 10% during the first 10 days of operation.

Assuming that the manganese removal in the test filters was achieved only by adsorption on MOCA, the filter with a 10 cm layer should display breakthrough after 24 hrs of filter operation (Table 6.6). However, no breakthrough was observed during more than a month of continuous operation of the test filters, most likely due to the presence of Birnessite and/or manganese oxidising bacteria in the fresh MOCA, promoting autocatalytic oxidation, resulting in the immediate formation of a new active MnO_x layer.

After about 2.5 days and in particular after 15 days of operation, a decrease in manganese removal efficiency was observed for the test filter with a 10 cm layer of MOCS. The reason was an operational malfunctioning, caused by a poor distribution of the feed water over the filter surface. Additionally, the feed water jet disrupted the top of the MOCA layer, locally thinning the layer to less than 10 cm. As soon as the MOCA layer was restored by a gentle backwash, manganese removal efficiency was re-established.

Based on the results depicted in Fig. 6.4(R), a 10 cm layer of fresh MOCA is sufficient to achieve an efficient manganese removal from the start of a new filter. However, to prevent practical problems (*e.g.*, short circuiting), it is recommended to apply a MOCA layer of at least 20 cm thickness.

Comparing the results of the full scale test filters with an added MOCS (Fig. 6.2^{B}) and MOCA layer (Fig. 6.2^{D-F}), a poor ripening of filters containing MOCS and a fast ripening of filters containing MOCA was observed. The major reason for the poor results achieved with the MOCS layer in a filter was probably, as explained above, attributed to the use of aged (dried) MOCS. The difference in performance could also be caused by the different feed water quality (*e.g.*, redox potential, NH_4^+ removal, pH), difference in MOCS and MOCA composition (coating Fe content) and different operational conditions applied (grain size of MOCS is approximately double that of MOCA, and Fe^{2+} loading and backwashing pattern and frequency were significantly different).

To summarise, this research showed that fresh manganese oxide-coated filter media were able to shorten the filter media ripening time substantially. Drying the MOCS has affected the results dramatically. In addition differences may have been caused by water quality as well as operational conditions. These phenomena must be investigated in more detail under comparable conditions, emphasising also the role of micro-biology and the importance and influence of specific bacteria.

6.5 Conclusions

The ripening time required to achieve complete manganese removal with (reference) full scale filters with virgin sand and virgin anthracite/sand filter media at two GWTPs; De Punt (The Netherlands) and Grobbendonk (Belgium) was found to be 55 and 16 days, respectively.

Differences in duration of ripening times between filters of the two GWTPs is caused by a combination of factors including the different composition of feed water (pH, redox potential, concentration of Fe^{2+} and NH_4^+), applied process design and operational conditions (*e.g.*, iron load, intensity and frequency of backwashing and physical properties and composition of virgin filter media).

Batch adsorption experiments demonstrated that both (fresh) MOCS and (fresh) MOCA adsorb Mn^{2+}. Based on Freundlich adsorption isotherm measurements, the manganese adsorption capacity (q_e) expressed per unit weight of adsorbent of MOCA was approximately twice the capacity of MOCS. However, when expressed per unit volume of adsorbent, which is more relevant for a practical application, the manganese adsorption capacities of MOCA and MOCS were similar. The adsorption capacity of commercial manganese adsorbent (Aquamandix), expressed per unit volume, was found to be approximately 3 times higher.

Aging and drying of MOCS, most probably resulted in the loss of manganese adsorption capacity. Besides, drying of MOCS may have caused the loss of auto-catalytic activity by changes in its structure and complete manganese oxidation. Finally the biological activity may have been lost.

The ripening time of a full scale filter with virgin anthracite/sand filter media, before reaching an effective manganese removal at GWTP Grobbendonk, of typically 16 days could be eliminated when a 0.10 to 0.30 m deep layer of fresh MOCA is placed on top of the virgin anthracite/sand filter bed. Because of operational aspects, it is advisable to apply a MOCA layer with a thickness ≥ 0.2 m.

In a follow up research, based on the results obtained from this study, the effect of water quality and operational conditions, as well as the role of micro-biology on filter media ripening, will be investigated in more detail under comparable conditions.

6.6 Acknowledgements

This research is financially and technically supported by WLN and the Dutch water companies Waterbedrijf Groningen and Waterleiding Maatschappij Drenthe. The authors would like to thank Mr A.A.S. Al Abri (MSc graduate at UNESCO-IHE) for his contribution to this work. Thanks also to the Belgian water company Pidpa, for providing a full scale test location Grobbendonk and their willingness to share the data from their groundwater treatment plant.

6.7 References

Appelo, C.A.J. & Postma, D. 2005. *Geochemistry, groundwater and pollution*, 2nd edition, CRC press, Boca Raton (FL), USA.

Bruins, J.H., Petrusevski, B., Slokar, Y.M., Kruithof, J.C. & Kennedy, M.D. 2014a. Manganese removal from groundwater: characterization of filter media coating. *Desalination and Water Treatment*, in press.

Bruins, J.H., Vries, D., Petrusevski, B., Slokar, Y.M. & Kennedy, M.D. 2014b. Assessment of manganese removal from over 100 groundwater treatment plants. *Journal of Water Supply: Research and Technology-AQUA,* 63(4), 268-280.

Buamah, R. 2009. Adsorptive removal of manganese, arsenic and iron from groundwater. *PhD thesis,* University Wageningen and UNESCO-IHE Delft, The Netherlands, ISBN 978-0-415-57379-5.

Buamah, R., Petrusevski B., de Ridder, D., van de Watering, T.S.C.M. & Schippers, J.C. 2009a. Manganese removal in groundwater treatment: practice, problems and probable solutions. *Water Science and Technology: Water Supply,* 9(1), 89-98.

Buamah, R., Petrusevski, B. & Schippers, J.C. 2008. Adsorptive removal of manganese (II) from the aqueous phase using iron oxide coated sand. *Journal of Water Supply: Research and Technology-AQUA,* 57(1), 1-11.

Burger, M.S., Mercer, S.S., Shupe, G.D. & Gagnon, G.A. 2008. Manganese removal during bench-scale biofiltration. *Water Research*, 42, 4733-4742.

Cools, B. 2010. Vlaamse Maatschappij voor Watervoorziening (VMW), Belgium. Personal communication.

Diem, D. & Stumm W. 1984. Is dissolved Mn^{2+} being oxidized by O_2 in absence of Mn-bacteria or surface catalysts. *Geochimica et Cosmochimica,* 48, 1571-1573

Fiol, N. & Villaescusa, I. 2009. Determination of sorbent point zero charge: usefulness in sorption studies. *Environmental Chemistry Letters,* 7, 79-84.

Flemming, H.C., Steele, H., Rott, U. & Meyer, C. 2004. Optimirung der in-situ reaktortechnologie zur dezentralen trinkwassergewinnung und grundwasseraufbereitung durch modelhafte untersuchungen beteiligter biofilme. *Report by the Institute for Sanitary Engineering,* Water Quality and Solid Waste Management of the University of Stuttgart.

Hu, P-Y., Hsieh, Y-H., Chen, J-C. & Chang, C-Y. 2004a. Adsorption of divalent manganese ion on manganese coated sand. *Journal of Water supply: Research and Technology – AQUA,* 53(3), 151-158.

Hu, P-Y., Hsieh, Y-H., Chen, J-C. & Chang, C-Y. 2004b. Characteristics of manganese-coated sand using SEM and EDAX analysis. *Journal of Colloid and Interface Science,* 272, 308-313.

Katsoyiannis, I.A. & Zouboulis, A.I. 2004. Biological treatment of Mn (II) and Fe (II) containing groundwater: kinetic considerations and product characterization. *Water Research,* 38, 1922-1932.

Kim, J. & S. Jung 2008. Soluble manganese removal by porous media filtration. *Environmental Technology,* 29(12), 1265-1273.

Kim, W.G., Kim, S.J., Lee, S.M. & Tiwari, D. 2009. Removal characteristics of manganese-coated solid samples for Mn (II). *Desalination and Water Treatment,* 4, 218-223.

Knocke, W.R., Van Benschoten, J.E., Kearny, M.J., Soborski, A.W. & Reckhow, D.A. 1991. Kinetics of manganese and iron oxidation by potassium permanganate and chlorine dioxide. *Journal of AWWA,* June, 80-87.

Krull, J. 2010. Stadtwerke Emden (SWE), Germany. Personal communication.
    NEN 2004 Water quality - Application of inductively coupled plasma mass spectrometry (ICP-MS) - Part 2: Determination of 62 elements, NEN-EN-ISO 17294-2.

Post, J.E. 1999. Manganese oxide minerals: Crystal structures and economic and environmental significance. *Proceedings of the National Academy of Sciences USA,* 96, 3447-3454.

Sahabi, D.M., Takeda, M., Suzuki, I. & Koizumi, J-I. 2009. Removal of Mn^{2+} from water by "aged" biofilter media: The role of catalytic oxides layers. *Journal of Bioscience and Bioengineering,* 107(2), 151-157.

Scherer, E. & Wichmann, K. 2000. Treatment of groundwater containing methane – combination of the processing stages desorption and filtration. *Acta Hydrochemica et Hydrobiologica,* 28(3), 145-154.

Stumm, W. & Morgan, J.J. 1996. *Aquatic Chemistry, Chemical Equilibria and Rates,* 3rd edition, Wiley, New York.

Tebo, B.M., Marger, J.R., Clement, B.G., Dick, G.J., Murray, K.J., Parker, D., Verity, R. & Webb, S.M. 2004. Biogenic manganese oxides: Properties and mechanisms of formation. *Annual Review of Earth and Planetary Sciences,* 32, 287-328.

Vandenabeele, J., De Beer, D., Germonpré, R. & Verstreate, W. 1992. Manganese oxidation by microbial consortia from sand filters. *Microbial Ecology,* 24, 91-108.

Vandenabeele, J., Van de Woestyne, M., Houwen, F., Germonpré, R., Vandesande, D. &Verstreate, W. 1995. Role of autotrophic nitrifiers in biological manganese removal from groundwater containing manganese an ammonium. *Microbial Ecology,* 28, 83-98.

Figure: *Filter backwashing (key factor controlling start of fast filter media ripening) at GWTP De Groeve (WBG) - (photo made by K. Borger, WLN, 2015)*

7 FACTORS CONTROLLING THE RIPENING OF MANGANESE REMOVAL FILTERS IN COVENTIONAL AERATION-FILTRATION GROUNDWATER TREATMENT

Main part of this chapter was submitted as:

Jantinus H. Bruins, Branislav Petrusevski, Yness M. Slokar, Koen Huysman, Koen Joris, Joop C. Kruithof, Maria D. Kennedy (2016). Factors controlling the ripening of manganese removal filters in conventional aeration-filtration groundwater treatment. Submitted to Desalination & Water Treatment.

7.1 Abstract

Relatively long operational time is required to achieve effective manganese removal in conventional aeration-filtration groundwater treatment with virgin filter media. Ripening period depends on water quality, operational parameters, and the filter media used. This study assessed the role of filter media type, backwashing and iron loading on the time required to achieve very effective manganese removal. Filter runs were conducted with two set-ups each with six pilot filters with virgin sand or anthracite, and different types of manganese oxide coated sand / anthracite (MOCS/MOCA). Pre-treated groundwater (aeration-rapid sand filtration), either directly, or after an additional pre-treatment (ultrafiltration-UF), was used as feed water. UF pre-treatment eliminated head loss development in pilot filters and backwashing was consequently not required. Filters that received feed water without UF pre-treatment required backwashing after 14 days of continuous operation. Use of virgin sand and anthracite resulted in comparable ripening time (25 days and 14 days for feed water without and with UF pre-treatment, respectively). Use of fresh MOCS/MOCA directly taken from operational filters, eliminated the need for ripening of virgin filter media, while dry MOCS was less effective than fresh one, while the total period required to achieve highly effective manganese removal (≥95%) was not shortened.

Keywords: filter backwashing, filter media ripening (manganese oxide coated sand/manganese oxide coated anthracite), groundwater quality, groundwater treatment, manganese removal.

7.2 Introduction

In many European countries the removal of manganese from groundwater is predominantly achieved by aeration-rapid sand filtration (Bruins *et al.*, 2014). This treatment approach requires no chemicals (to oxidize Mn^{2+}), in contrast to manganese removal based on oxidation-filtration that is commonly applied in US and some other countries (Knocke *et al.*, 1991), is easy to operate, and very cost-effective. Application of this process, is, however, associated with a number of challenges including the long ripening period of virgin filter media required to achieve very effective manganese removal (Buamah *et al.*, 2009; Cools, 2010; Krull, 2010). Several parameters are suggested to influence virgin filter media ripening period, including groundwater quality, type of filter media, and intensity and frequency of filter backwashing (Bruins *et al.*, 2014; Paassen van, 2010).

Removal of manganese (Mn^{2+}) in aeration-rapid sand filtration, which is often believed to be an auto-catalytic adsorption-oxidation process (Graveland and Heertjes, 1975; Stumm and Morgan, 1996), can be supported by biological manganese oxidation (Vandenabeele *et al.*, 1992; Tebo *et al.*, 2004; Barger *et al.*, 2009; Burger *et al.*, 2008). Formation of manganese oxide (MnO_x) is likely initiated biologically, and over a prolonged filter run time, MnO_x becomes of predominantly physico-chemical origin (Bruins *et al.*, 2015a).

Many researchers have shown that the removal of dissolved manganese (Mn^{2+}), can be facilitated by adsorption on manganese oxide coated filter media (Hu *et al.*, 2004a,b; Kim and Jung, 2008; Kim *et al.*, 2009). Filter media suggested to be the most efficient is manganese oxide coated sand / anthracite – MOCS / MOCA (Olanczuk-Neyman and Bray, 2000; Stembal *et al.*, 2005; Tiwari *et al.*, 2007; Buamah *et al.*, 2008; Tekerlekopoulou and Vayenas, 2008; Sahabi *et al.*, 2009; Islam *et al.*, 2010; Bruins *et al.*, 2015b). Addition of a layer of fresh MOCA in the filter with virgin anthracite was reported to completely eliminate filter media ripening time, while at an another testing location provision of a layer of dried MOCS did not result in any reduction (Bruins *et al.*, 2015c). It was hypothesized that differences in MOCS / MOCA performance at two testing locations could be explained by dissimilar conditions, under which the tests were carried out, such as water quality, mode of backwashing and different characteristics of manganese oxide coated media.

The aim of this research was to study the effect of backwashing, and, the type of virgin filter media on the duration of the ripening period required to achieve very effective (> 90%) manganese removal. In addition, the potential of freshly taken manganese coated filter media, from very effective operating manganese removal filters (MOCS as well as MOCA), to reduce ripening period of manganese removal filters with virgin filter media was investigated. Also dried manganese coated media (with a potential loss of biological activity and adsorptive properties) were tested.

7.3 Materials and Methods

Manganese oxide coated materials used in this study were obtained from three full scale aeration-filtration groundwater treatment plants (GWTPs). Figure 7.1 gives an overview of the process schemes of the GWTPs, and the filters from which Manganese Oxide Coated Sand (MOCS) / Manganese Oxide Coated Anthracite (MOCA) samples were taken.

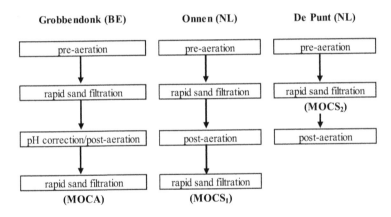

Figure 7.1: Treatment schemes of GWTPs and filters from which MOCS/MOCA samples were taken.

Table 7.1 gives an overview of quality of feed water to filters, from which MOCS and MOCA were taken.

Table 7.1:*Quality of feed water to filters, from which MOCS samples were taken*

Parameter	Unit	Grobbendonk	Onnen	De Punt
Iron	mg/L	0.09 (0.03 - 0.1)	0.11 (0.02 - 0.5)	5.9 (5.2 - 6.4)
Manganese	mg/L	0.13 (0.10 - 0.15)	0.06 (0.03 - 0.15)	0.23 (0.19 - 0.27)
Ammonium	mg/L	0.08 (< 0.02 - 0.20)	0.06 (0.05 - 0.09)	0.46 (0.34 - 0.61)
pH	[-]	7.6 (7.7 - 7.9)*	7.5 (7.4 - 7.6)	7.1 (7.0 - 7.1)
Oxygen	mg/L	9 (8 - 9.5)	8 (6.8 - 9.1)	10 (9.3 - 10.6)

* *pH mainly found to be between 7.5 and 7.6*

Fresh MOCA and MOCS$_1$ were taken from well operating filters (manganese removal ≥95%), and directly (without drying) used in pilot filter columns. In addition, two batches of dried MOCS$_1$ and MOCS$_2$ (taken out of manganese removal filters and subsequently dried in open air, and stored for 2 and 6 months, respectively), were also used in this research. Dried coated filter media was included in the research having in mind that it is not always possible to obtain fresh manganese coated media that can be used for a start-up of manganese removal filters with virgin filter media. Prior to using, all filter media samples were rinsed, to flush out fines.

The pilot filters operated at GWTP Grobbendonk (Belgium). The experimental set-up (Fig. 7.2) consisted of two sets each with 6-columns installed in a parallel.

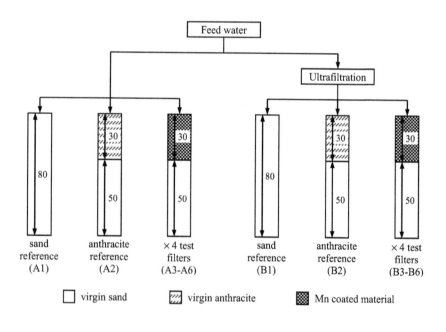

Figure 7.2: *Pilot set up at GWTP Grobbendonk*

Based on full scale experiments in Grobbendonk, a filter media layer of 30 cm was used. The pilot filter columns had a diameter of 10 cm, and the filter bed length was 80 cm, composed of 50 cm support material (virgin sand) and 30 cm virgin sand, anthracite, MOCA or MOCS as follows:

- A1 / B1: virgin sand (reference for MOCS layers);
- A2 / B2: virgin anthracite (reference for MOCA layers);
- A3 / B3: MOCA (fresh);
- A4 / B4: MOCS₁ (fresh);
- A5 / B5: MOCS₁-dried;
- A6 / B6: MOCS₂-dried.

The physical properties of MOCS and virgin sand/anthracite, the chemical composition of the coating and the Freundlich adsorption isotherm constants for Mn^{2+} adsorption on MOCS₁, MOCS₂ and MOCA (all fresh) are given in Table 7.2.

Supernatant layer of 0.3 m was provided above filter media. The pilot filter columns were operated in down flow mode at constant filtration rate of 5.1 ± 0.5 m/h. The columns in test set A were backwashed after approximately every 2 weeks of continuous operation. Backwashing was carried out with water only, at a backwash rate of 30-35 m/h, resulting in approximately 10-20 % filter bed expansion, with typical duration of 10 min. The columns in the test set B, that received water after additional UF pre-treatment, were not backwashed during the whole study period, because no increase of head loss was observed

Table 7.2: *Physical properties of fresh MOCS, MOCA, and virgin sand / anthracite, chemical composition of the coating and the Freundlich adsorption isotherm constants for Mn^{2+} adsorption on fresh MOCS and MOCA*

Parameter	Unit	MOCS₁	MOCS₂	MOCA	Virgin sand	Virgin anthracite
Grain size (d_{10}-d_{90})	mm	1.3 - 2.0	1.6 - 3.1	0.8 - 1.5	0.4 - 0.8	0.8 - 1.6
Uniformity coefficient	[-]	1.21	1.58	1.52	1.30	1.42
Bulk density	Kg/L	1.376	1.177	0.650	1.459	0.635
Particle density	Kg/L	2.332	2.326	1.176	2.586	1.400
Porosity	%	46.0	49.4	44.7	43.6	54.6
pH$_{PZC}$	[-]	7.8	7.2	8.0	6.1 - 6.5	9.1
Coating composition						
Mn	mg/g	30.4	12.8	13.5	-	-
Fe	mg/g	21.4	158	2.2	-	-
Ca	mg/g	7.8	8.9	2.5	-	-
Al	mg/g	2.4	0.5	0.3	-	-
Freundlich adsorption isotherm constants						
K	(mg/g)/(mg/L)	0.70	0.45	0.91	-	-
1/n	[-]	1.25	1.31	1.34	-	-
r^2	[-]	0.97	0.91	0.91	-	-
q_e at C_e = 0.2 mg/L	mg/g	0.193	0.132	0.276	-	-
q_e at C_e = 0.2 mg/L	g/L	0.265	0.155	0.179	-	-

The filtrate from the first stage of the full scale GWTP Grobbendonk was used directly, or after ultrafiltration -UF (Inge, dizzer 500SB, pore size 0.02 µm), as feed for the pilot filter columns. UF filtration was applied to retain particles present in the feed water. It was assumed that ultrafiltration would prevent, or at least strongly reduce, head loss development in the filter bed, with associated reduced backwashing frequency. Filtrate of the first filtration step is also feed to full scale manganese removal filters at this plant (Table 7.1). Composition of pilot feed water is given in Table 7.3.

Manganese in feed water was in dissolved (Mn^{2+}) form, while iron was present predominantly as (hydr)oxide.

Samples of feed water and filtrate (a sampling point just below the top layer of 30 cm), were taken daily, and the manganese concentration was analysed with ICP-MS.

Table 7.3: *Quality (average and range) of feed water to pilot filters, with and without UF pre-treatment*

Parameter	Unit	Without UF	With UF
Iron	mg/L	0.09 (0.03 - 0.1)	< 0.02
DOC	mg/L	1.58 (1.51 - 1.68)	1.46 (1.33 - 1.60)
Manganese	mg/L	0.13 (0.10 - 0.15)	
Ammonium	mg/L	0.08 (< 0.02 - 0.20)	
pH	[-]	7.6 (7.5 - 7.9)*)	
Oxygen	mg/L	9 (8 - 9.5)	
Redox potential	mV	+250 (+200 - +290)	

*) *most of the time pH ranged from 7.5 to 7.6*

7.4 Results and Discussion

7.4.1 Filter ripening with virgin media and the effect of filter backwashing

Fig. 7.3 shows the comparison of virgin sand and virgin anthracite ripening time, and the effect of UF pre-treatment on the ripening period during the first 25 days of filter run with virgin filter media.

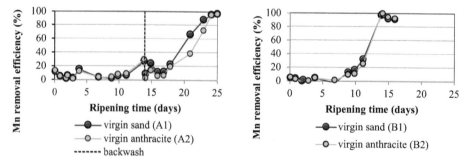

Figure 7.3: *Mn removal efficiency (%) as a function of the filter ripening time, type of virgin filter media and UF pre-treatment; feed water without (left), and with UF pre-treatment (right)*

From Fig. 7.3 it can be seen that ripening time required to achieve effective manganese removal with virgin sand and anthracite, was similar when the same feed water pre-treatment was applied. The results imply that

112

different physical properties of sand and anthracite (*e.g.* particle size and pH_{PZC}, see Table 7.2.) did not significantly affect Mn^{2+} adsorption, and related rate of Mn-coating development. However, UF pre-treatment of feed water (Fig. 7.3 right) resulted in faster filter media ripening. Highly effective manganese removal ($\geq 95\%$) was achieved after approximately 14 days, as compared to 24 days in the filters when no UF was applied. The difference in media ripening time of the two set-ups could be presumably attributed to the effect of backwashing on development of the Mn-coating (Bruins *et al.*, 2015a) . Mn-coating is known to be essential for effective manganese removal in aeration-rapid sand filtration treatment (Hu *et al.*, 2004a,b; Kim and Jung, 2008; Kim *et al.*, 2009). In general, it is known that the oxidation-reduction potential (ORP) influences the manganese removal efficiency. However, in this research the influence of ORP was limited. During the whole test, ORP was close to around + 250 mV. Also after each backwashing.

During the first two weeks no filter backwash was applied (head loss developed in filter bed was compensated by opening and adjusting a flow control valve to allow operation of pilot column at constant filtration rate) and the ripening of all filters with virgin filter media in both set-ups was very similar, irrespective of applied pre-treatment. Backwashing of filters A1 and A2 conducted after 14 days of continuous operation, however, resulted in reduction of the Mn^{2+} removal efficiency from 28 % to <10 % (Fig 7.3 left). Due to the UF pre-treatment, the feed water of columns B1 and B2 contained no particulate matter and as a consequence no filter backwashing of columns B1 and B2 was required (Fig 7.3 right). Although the backwashing used in this study was gentle and performed only after 14 days (Fig. 7.3 left, fine dotted line), the observed differences in Mn removal efficiency suggest that backwashing has a substantial effect on the ripening time of filter media.

Negative effect of backwashing during the ripening phase on manganese removal could be likely attributed to:

- (partial) removal of bacteria responsible for oxidation of Mn^{2+} that developed on virgin filter media during ripening period (Vandenabeele *et al.*, 1992; Tebo *et al.*, 2004; Barger *et al.*, 2009; Burger *et al.*, 2008; Bruins *et al.*, 2015a) and / or
- (partial) removal of freshly formed MnO_X (Birnessite), which has highly autocatalytic properties to adsorb and subsequently oxidize Mn^{2+} (Bruins *et al.*, 2015b; Post, 1999).

Fig. 7.4 shows the effect of the filter backwashing on manganese removal during the total testing period of approximately 70 days. Manganese concentration in filtrate was analysed before and after each backwash cycle.

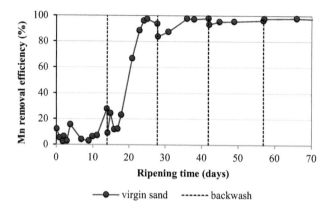

Figure 7.4: *Effect of filter backwashing on manganese removal efficiency during filter media ripening with virgin sand.*

From Fig. 7.4 it can be seen that directly after each backwash cycle, the manganese removal efficiency decreased. The decrease was most pronounced after the first backwash cycle (manganese removal reduced from 28 % to <10 %), carried out 14 days after the start of the filter operation, when the manganese removal was still rather ineffective presumably due to only limited partial media coating with birnessite (Bruins *et al.,* 2015b). Backwashing at that stage of ripening period likely partially removed initial birnessite deposits, and bacteria responsible for Mn^{2+} oxidation, and, consequently had a strong negative impact on the Mn removal efficiency. After the second and the third backwash cycles (performed after 28 and 42 days of filter operation), the reduction of manganese removal efficiency was less pronounced, but still obvious - from 94 % to 84 % after the second backwash cycle, and from 98 % to 93 % after the third backwash cycle. After the fourth backwash cycle (57 days of the filter operation), no significant decrease of manganese removal efficiency was observed, showing that there was sufficient birnessite on filter media, and partial removal due to backwashing did not significantly hampered manganese removal.

Filters treating iron and manganese containing groundwater need to be periodically backwashed to remove particles caught in filter bed voids, causing head loss development. The majority of these particles are iron (hydr) oxides formed by oxidation of dissolved iron with oxygen. The backwashing of pilot filters in this study was carried out with a very low frequency (approximately once in 14 days) due to low iron concentration in the feed water (up to a maximum of 0.1 mg/L for the set-up A), and as a consequence, iron loading in the columns was limited (approximately 0.1 kg Fe/m² filter area, per filter run). In practice, however, iron loading can be much higher, introducing the need for more frequent backwash cycles (*e.g.,* the backwash frequency of GWTP Noordbargeres, water company Drenthe (NL), is more than once per day, because of high iron concentration in the feed water of approximately 14 mg/L, with iron loading of 1.35 kg Fe/m²/filter run). Under such conditions, the negative effect of backwashing on duration of ripening period required to achieve very effective manganese removal with virgin filter media will be much more pronounced (very effective manganese removal in the filter is typically achieved after more than four months).

In Fig. 7.5 the filter media ripening time of the pilot filter from this study was compared with the filter ripening time of the full scale filter GWTP Grobbendonk Bruins *et al.,* 2015c). In both cases the virgin anthracite was used as filter media. Feed water quality for both filters was the same (Table 7.3, column "without UF"). However, the applied filtration rate, and consequently iron loading and backwash frequency were, different. In addition, the depth of the anthracite layer was different (0.5 m and 0.3 m in the full scale and the pilot filters, respectively). Inspection of the anthracite layer after completion of filter run showed that the manganese was removed mainly in the top 0.10-0.15 m of the filter, suggesting that different heights of the anthracite layer did not affect the results.

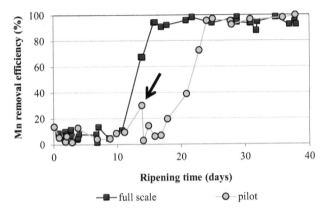

Figure 7.5: *Comparison of the Mn removal efficiency as a function of the filter ripening time for full scale and pilot filters with virgin anthracite (filtration rate: Vf = 5.1 m/h pilot filter and Vf = 2.5 m/h full scale filter).*

Highly effective manganese removal (≥ 95%) was achieved after approximately 16 days in the full scale plant, and after 24 days in the pilot filter column. The difference in applied filtration rate (2.5 m/h and 5.1 m/h, for full scale and pilot filters, respectively), resulted in different iron loadings (0.05 and 0.1 kg Fe/m² filter area, per filter run). As a consequence, the pilot filter column had to be backwashed earlier, than the full scale filter (14 and 30 days, respectively). The first backwashing of the pilot column took place just when manganese removal efficiency started to rapidly increase (black arrow in Fig. 7.5), while for the full scale filter, the first backwashing was applied only after 30 days, when the highly effective manganese removal was already achieved.

Iron concentrations in feed water of drinking water treatments plants may range from < 0.02 mg/L to more than 30 mg/L, with related iron loading from less than 0.01 to over 10 kg/m² filter area, per filter run (Bruins *et al.,* 2014). As a consequence, the backwash frequency in practice can vary between once per month to more than once per day. Findings emerging from this study, suggest that the ripening time required to achieve very effective manganese removal with virgin filter media is strongly affected by the applied backwash frequency that is correlated to the iron loading. High iron loading, and consequently high backwash frequency, could be reduced by operating filters with lower filtration rate during the ripening period. Another option is to recirculate part of the filtrate, thus lowering the iron loading and consequently the backwash frequency. The reduction of the plant operational capacity during the ripening period will in most cases not be a problem because water with high manganese concentration, has to be disposed anyway. Backwashing with water only is recommended during the ripening period to limit removal of bacteria and MnOₓ deposits developed on the filter media, since backwashing with air and water, which is normally used, is much more abrasive.

7.4.2 Filter ripening with addition of a layer of manganese coated media

Fig. 7.6 shows the manganese removal efficiency of pilot filters with virgin sand containing a layer of fresh MOCA or $MOCS_1$ during the initial 2-3 weeks of operation.

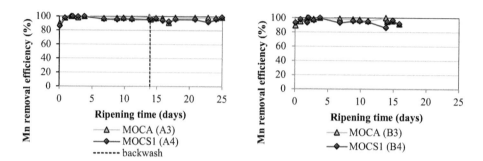

Figure 7.6: *Mn removal efficiency (%) as a function of the filter run time for pilot filters containing a layer of fresh $MOCS_1$ or MOCA; feed water without (left) and with UF (right) pre-treatment.*

Fig. 7.6 shows that the addition of a layer of either fresh MOCA or $MOCS_1$ on top of the virgin sand resulted in very effective (>90 %) manganese removal already after one day of filter operation, irrespective of UF pre-treatment. It has been shown, that birnessite presence in filter media coating is essential for an effective removal of dissolved manganese (Bruins *et al.,* 2015b). The results from this study show that MOCS or MOCA, freshly taken from well performing manganese removal filters were auto-catalytically active. This assumption was supported by calculation of the maximum adsorption capacity of the MOCS/MOCA layers (2.35 litres). Based on the adsorption isotherms (Table 7.2) Mn^{2+} adsorption capacity (at Ce=0.2 mg/L) was calculated to be 623, 365 and 420 mg Mn^{2+} for $MOCS_1$, $MOCS_2$ and MOCA, respectively. Given the daily filtrated volume through each column (960L) and the average Mn^{2+} concentration in the feed water (0.13mg/L) the adsorption capacity of MOCS layers expected to be exhausted after 5.0; 2.9 and 3.4 days, for $MOCS_1$, $MOCS_2$ and MOCA, respectively. Fig. 6 shows that manganese removal was still very effective after 25 days of continuous pilot filters operation. This strongly suggests the presence of birnessite, that enables an efficient auto-catalytic adsorption and subsequent oxidation of Mn^{2+} (Bruins *et al.,* 2015b; Post, 1999). Therefore replacing the top layer of a new (virgin) filter by 'active' manganese coated media could eliminate the ripening time of manganese removal filters.

Fig. 7.6 also shows that the manganese removal efficiency of fresh MOCA and $MOCS_1$ was almost identical, as expected based on the Freundlich adsorption isotherm constants (Table 7.2). Use of fresh MOCA or MOCS, from well performing manganese filters (even with different physical properties and chemical composition) can consequently reduce, or even eliminate the long filter ripening period.
Results obtained (Fig 7.6 left) also confirmed that backwashing has rather limited effect on manganese removal for filters media with well developed (birnessite) coating.

Fig. 7.7 shows the manganese removal during initial 2-3 weeks of operation of pilot filter columns with virgin sand and a layer of dry $MOCS_1$ and $MOCS_2$, or a layer of fresh $MOCS_1$. Feed water without (Fig. 7.7, left) and with UF pre-treatment (Fig. 7.7, right) was used in these filter runs.

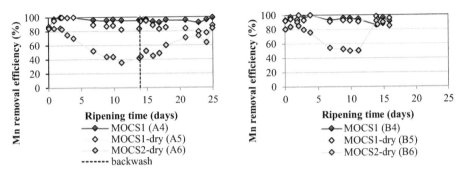

Figure 7.7: *The Mn removal efficiency as a function of filter run time for pilot filters containing layers of dried or fresh MOCS; feed water without (left) and with UF (right) pre-treatment*

Fig. 7.7 shows that the performance of the filters containing a layer of dry manganese coated media was not as efficient as the filters with a layer of a fresh manganese coated media. The difference was more pronounced when no UF pre-treatment was applied (Fig 7.7 left). Compared to the fresh material, the removal of manganese was less efficient from the start of the filter run of the pilot filters with a layer of MOCS2-dry, while for pilot filters with MOCS1-dry showed only during the initial 5 days of operation very effective Mn removal (>95%), that was subsequently reduced to approximately 80% and 90% for feed water without and with UF pre-treatment, respectively. For MOCS2-dry the Mn removal efficiency started strongly decreasing from the 3rd day of filter run.

Dry MOCS is less effective for manganese removal due to several reasons. Firstly, MOCS during drying process may have lost part of its adsorptive capacity, presumably due to oxidation of autocatalytic active MnO_x (birnessite) into less autocatalytic active MnO_x (pyrolusite). As a result, the number of available adsorptive sites decreases (Post, 1999). Secondly, drying the MOCS can have negative effect on the structure of the manganese oxide. The auto-catalytically active birnessite, consisting of plates (Post, 1999), may irreversibly collapse during drying, subsequently additionally limiting the number of available adsorption sites. Finally, filter media drying might cause a loss of the biological activity on the media surface due to manganese oxidizing bacteria die-off. These bacteria likely play an important role in the process of manganese oxidation and removal (Vandenabeele *et al.*, 1992; Tebo *et al.*, 2004; Barger *et al.*, 2009; Burger *et al.*, 2008; Bruins *et al.*, 2015a).

Fig. 7.7 also shows that dry MOCS2 removed manganese less effectively than dry MOCS1-dry. The differences between the two MOCS media could be likely attributed to difference in manganese content and associated available adsorptive sites (*e.g.* MOCS2-dry that was less effective for manganese removal contained 12.8 mg Mn/g, while MOCS1-dry contained 30.4 mg Mn/g). Another explanation for the differences in removal capacities of the two dry MOCS media might be the storage time of two materials; MOCS1-dry was stored for 2 months, whereas MOCS2-dry was stored for 6 months before their use in this study. Longer storage (in a dry air) presumably resulted in more pronounced loss of auto-catalytic properties through mechanisms earlier discussed.

Backwashing of the pilot filter with a layer of dry MOCS1 and MOCS2 had a short positive effect on manganese removal that could be likely explained by removal of iron precipitates that physically blocked adsorption sites on filter media.

In both pilot filters with dry manganese coated media, irrespective of the UF pre-treatment, manganese removal efficiency started to improve after 12 days of pilot filter operation. At approximately the same time manganese removal started increasing in pilot filters with virgin sand or anthracite (Fig. 7.3). The coinciding ripening times indicate that dry manganese coated media behaved similar to virgin filter media, likely due to formation of manganese oxides and a biologically active layer that were required to facilitate effective manganese removal.

The findings of this research can be used in practice: addition of a layer ($\geq$0.3m) of fresh manganese coated media, strongly reduces or completely eliminates the filter ripening time. Manganese coated media can easily be taken from a well-operating manganese removal filter and transferred to a filter with virgin media. This procedure has to be done only once, and there is no need for follow up replacements of manganese coated filter media.

Filters are immediately ready for operating after using a fresh layer of MOCS or MOCA. As a consequence of this fast filter media ripening procedure, a water company saves operational costs (saving water, labor, costs for analysis, no need for extra filters, etc.). Savings are strongly depending on the original ripening time, without the use of manganese coated filter media.

If no fresh layer of manganese oxide coated filter media from a well-operating manganese removal filter can be used in practice, the long ripening period of manganese removal filters with virgin filter media can be reduced by operating filters, temporarily, at a lower filtration rate during the ripening period, consequently reducing the iron loading and thus limiting the backwashing frequency. Although in this situation the water production capacity of the filter is, temporarily, also reduced, possible advantages are more pronounced.

7.5 Conclusions

Findings emerging from this study focused on the investigation of key factors controlling the ripening period of manganese removal filters with virgin media in conventional aeration-filtration treatment lead to the following conclusions:

- Despite different physical properties, virgin sand and virgin anthracite will have similar ripening time required to achieve very effective manganese removal, assuming that the feed water quality and operational conditions are identical.
- Filter backwashing prolongs the ripening time of a virgin filter. The influence of backwashing becomes less pronounced with the progress of filter media coating, due to development of thicker layer of biomass and / or auto-catalytically active birnessite on adsorption media surface.
- Backwashing has no, or only limited influence, on manganese removal when a layer of fresh manganese oxide coated filter media from well performing manganese removal filters, is added to a filter bed with virgin media.
- Ripening of manganese removal filters with virgin filter media can be shortened by temporarily operating filters at lower filtration rate with associated less frequent backwashing cycles.
- Addition of a layer ($\geq$0.3m) of fresh manganese coated media, from a well-operating manganese removal filter, to a filter with virgin media, strongly reduces or completely eliminates the filter ripening time.
- Compared to fresh-, dry MOCS has inferior manganese removal properties due to lower adsorption capacity and likely very limited potential for catalytic Mn^{2+} adsorption-oxidation.

- Time required to create a new (bio-active) auto-catalytically layer of birnessite on the surface of dry manganese coated or virgin filter media is similar.
- The long ripening period of manganese removal filters with virgin filter media in practice, can be reduced or even eliminated by (1) addition of a layer of fresh MOCA or MOCS from a well-functioning manganese removal filters, and (2) operating filters at lower filtration rate during the ripening period consequently reducing the iron loading and the backwashing frequency.

7.6 Acknowledgements

This research was financially and technically supported by WLN and the Dutch water companies Groningen (WBG) en Drenthe (WMD). The authors wish to thank water company Pidpa (Belgium), for their willingness to share the data from their groundwater treatment plant and for the opportunity to carry out pilot tests at their Grobbendonk facilities. Special thanks are extended to Mrs. Ann Maeyninckx and Mrs. Martine Cuypers (Pidpa) for helping run the pilot and performing countless analyses.

7.7 References

Barger J.R., Fuller C.C., Marcu M.A., Brearly A., Perez De la Rosa M., Webb S.M., Caldwell W. A. 2009. Structural characterization of terrestrial microbial Mn oxides from Pinal Ckeek. AZ. Ceochimica et Cosmochimica Acta, 73, 889-910.

Bruins J. H., Vries D., Petrusevski B., Slokar Y.M., Kennedy M.D. 2014. Assessment of manganese removal from over 100 groundwater treatment plants. Journal of Water Supply: Research and Technology – AQUA, 63 (4) 268-280.

Bruins J.H., Petrusevski B., Slokar Y.M., Huysman K., Joris K., Kruithof J.C., Kennedy M.D. 2015a. Biological and physico-chemical formation of Birnessite during the ripening of manganese removal filter, Water Res. 69, 154-161.

Bruins J.H., Petrusevski B., Slokar Y.M., Kruithof J.C., Kennedy M.D. 2015b. Manganese removal from groundwater: characterization of filter media coating. Desalination & Water Treatment, 55 (7), 1851 - 1863.

Bruins J.H., Petrusevski B., Slokar Y.M., Huysman K., Joris K., Kruithof J.C., Kennedy M.D. 2015c. Reduction of ripening time of full scale manganese removal filters with manganese oxide coated filter media. Journal of Water Supply: Research and Technology – AQUA, 64 (4), 434-441.

Buamah R., Petrusevski B., Schippers J.C. 2008. Adsorptive removal of manganese (II) from the aqueous phase using iron oxide coated sand. Journal of Water supply: Research and Technology – AQUA, 57 (1), 1-11.

Buamah R., Petrusevski B., de Ridder D., van de Watering S., Schippers J.C. 2009. Manganese removal in groundwater treatment: practice, problems and probable solutions. Journal of Water Science and Technology: Water Supply, 9 (1) (2009) 89-98.

Burger M.S., Krentz C.A., Mercer S.S., Gagnon G.A. 2008. Manganese removal and occurrence of mangnanese oxidizing bacteria in full-scale biofilters, J. Water Supply Res. Technol. AQUA 57.5. 351 - 359.

Cools B. 2010. De Watergroep (Flemish water company), Personal communication, Belgium.

Graveland A., Heertjes P.M. 1975. Removal of mangnanese from groundwater by heterogeneous autocatalytic oxidation. Trans. Chem. Eng. 53, 154-164.

Hu P-Y., Hsieh Y-H., Chen J-C., Chang C-Y 2004a. Adsorption of divalent manganese ion on manganese coated sand. Journal of Water Supply: Research and Technology – AQUA, 53 (3), 151-158

Hu P-Y., Hsieh Y-H., Chen J-C., Chang C-Y. 2004b. Characteristics of manganese-coated sand using SEM and EDAX analysis. Journal of Colloid and interface science, 272, 308-313.

Islam A.A., Goodwill J.E., Bouchard R., Tobiasen J.E., Knocke W.R.. 2010. Characterization of filter media $MnO_2(s)$ surfaces and Mn removal capability. Journal AWWA, 102 (9), 71-83.

Kim J., Jung S. 2008. Soluble manganese removal by porous media filtration. Environmental Technology, 29 (12), 1265-1273.

Kim W.G., Kim S.J., Lee S.M., Tiwari D. 2009. Removal characteristics of manganese-coated solid samples for Mn(II). Desalination and Water Treatment, 4, 218-223.

Knocke W.R., van Benschoten J.E., Kearny M.J., Soborski A.W., Reckhow D.A. 1991. Kinetics of manganese and iron oxidation by potassium permanganate and chlorine dioxide. Journal of AWWA June, 80-87.

Krull J. 2010. Stadwerke Emden – SWE (German water company), Personal communication, Germany.

Olanczuk-Neyman K., Bray R. 2000. The role of physico-chemical and biological processes in manganese and ammonia nitrogen removal from groundwater. J. Pol Environ. Stud., 9 (2), 91-96.

Paassen van J. 2010. Vitens (Dutch water company), Personal communication, The Netherlands.

Post J.E. 1999. Manganese oxide minerals: Crystal structures and economic and environmental significance. Proceedings of the National Academy of Sciences USA, Vol 96, 3447-3454.

Sahabi D.M., Takeda M., Suzuki I., Koizumi J-I. 2009. Removal of Mn^{2+} from water by "aged" biofilter media: The role of catalytic oxides layers. Journal of Bioscience and Bioengineering, 107 (2), 151-157.

Stembal T., Markic M., Ribicic N., Briski F., Sipos L. 2005. Removal of ammonia, iron and manganese from ground waters of Northern Croatia – pilot plant studies. Process Biochemistry, 40, 327-335.

Stumm, W., Morgan J.J. 1996. Aquatic chemistry, Chemical Equilibria and Rates, third ed. Wiley, New York.

Tebo B.M., Marger J.R., Clement B.G., Dick G.J., Murray K.J., Parker D., Verity R., Webb S.M 2004. Biogenic Manganese oxides: Properties and mechanisms of formation. Annu. Rev. Earth Planet Sci., 32, 287-328.

Tekerlekopoulou A.G., Vayenas D.V. 2008. Simultaneous biological removal of ammonia, Iron and manganese from potable water using a trickling filter. Biochemical Engineering Journal, 39, 215-220.

Tiwari D., Yu M.R., Kim M.N., Lee S.M., Kwon O.H., Choi K.M., Lim G.J., Yang J.K. 2007. Potential application of manganese coated sand in the removal of Mn (II) from aqueous solutions. Water Science & Technology, 56 (7), 153-160.

Vandenabeele J., de Beer D., Germonpré R., Verstreate W. 1992. Manganese oxidation by Microbial consortia from sand filters. Microbial Ecology, 24 (1992) 91-108.

Figure: *Pilot set up at GWTP De Grobbendonk (water company Pidpa) - (photo made by J. H. Bruins, WLN, 2012)*

8 GENERAL CONLUSIONS

8.1 Overall conclusions and perspective

The main challenge, regarding manganese removal from groundwater with aeration-rapid sand filtration, is to get a better understanding of the mechanisms involved in the ripening of virgin filter media and to use this knowledge to shorten or preferably completely eliminate the long ripening period of virgin media and to prolong the lifetime of the filter media.

Based on the research described in this thesis it can be concluded that several water quality and operational parameters, such as iron (and iron loading), pH, ammonia and empty bed contact time are of importance for efficient manganese removal. Furthermore, *Birnessite* was identified as the oxide predominantly involved in effective manganese removal. Birnessite is highly reactive, with outstanding auto-catalytic adsorptive and oxidative properties. This research clearly demonstrates (at pilot scale) that Birnessite was produced *biologically* at the start of filter ripening but as filter ripening progressed, the production of Birnessite was *physico-chemical*. During this study, molecular DNA analysis showed a bacteria population shift from the iron oxidizing genus *Gallionella*, found in iron removal filters, to the manganese and nitrite oxidizing genus *Pseudomonas* and *Nitrospira*, present in the manganese removal filter. Furthermore, bacteria species that could potentially oxidise manganese were also identified.

During pilot research, key factors, for efficient start-up of manganese removal filters were defined. Most important is the influence of backwash frequency on manganese removal efficiency at the start of filter media ripening. Furthermore it was found that the use of freshly prepared manganese oxide coated media practically eliminated the ripening period. Optimal conditions to shorten filter ripening (in practice) using MOCS and MOCA, containing Birnessite, were established.

The findings and conclusions of this research defined important water quality and operational parameters and knowledge about the role bacteria play in rapidly ripening filter media. Knowing the bacteria involved and being aware of the damaging effect of filter backwashing, especially at the start of filter ripening, yields a solid base for a better understanding of the mechanisms involved in speeding up the ripening process with virgin filter media. This information provides water companies the opportunity to develop an *innovative* manganese removal process *i.e.,* 'a manganese oxidizing bacteria-friendly process', with a restricted backwash frequency at the start-up of filter ripening. And if possible using fresh MOCS or MOCA to shorten or preferably completely eliminate the long ripening period of new filters and prolong the useful lifetime of the media.

In this chapter each individual aim is discussed in more detail in a separate section to place the objectives in a broader perspective. Finally a general outlook with some recommendations for use in practice and further research are given.

8.2 Assessment of manganese removal from over 100 groundwater treatment plants

The aim of this study was to make an inventory of water quality and operational parameters affecting manganese removal through aeration-rapid sand filtration and to establish correlations between these parameters and manganese removal efficiency.

The focus of the overview was on the manganese removal efficiency in the first filtration stage. Data from over 100 selected full-scale groundwater treatment plants were collected, and univariate and multivariate statistical (PCA) analyses conducted. Furthermore, close inspection of the collected data and conducted statistical analyses (univariate correlations and PCA) showed that the manganese removal efficiency was influenced by several parameters simultaneously, including both water quality and operational (design) parameters.

The multivariate statistical method (PCA) revealed that iron loading, NH_4^+ removal efficiency and pH of the filtrate played a major role in manganese removal, while oxygen concentration in the filtrate, and phosphate, manganese and NH_4^+ concentration in the raw water influenced manganese removal to a lower degree. Operational parameters, such as filtration rate and empty bed contact time (EBCT) were found to be of secondary importance.

PCA, Univariate statistics and assessment of available data indicated that a very effective manganese removal efficiency in the first aeration-filtration stage with simultaneous removal of iron and ammonia, was achieved under the following conditions:

- NH_4^+ removal efficiency : $> 85\%$
- iron loading per filter run : < 2.7 kg Fe/m^2
- pH of filtrate : > 7.1
- filtration rate : < 10.5 m/h
- empty bed contact time : > 11.5 min
- oxygen in filtrate : ≥ 1 mg/L

The results of this study give directions to engineers to design a traditional aeration-filtration process for efficient manganese removal. Based on the raw water quality to be treated, and the results of this study the most suitable filter configuration and the required filter dimensions and filtration conditions can be selected. In addition, results emerging from univariate statistics and assessment of available data from full scale plants can help in making a decision if manganese removal can be combined with removal of iron and ammonium in a single filtration step. Therefore, the amounts of iron, ammonia and manganese to be treated must be known and based on the conditions at which they can be removed together in one filtration step, provided by this study, it can be decided if application of one filtration step only is feasible for a new treatment design.

8.3 Manganese removal from groundwater: Characterization of filter media coating

Removal of manganese in conventional aeration-filtration groundwater treatment plants (GWTPs) results in the formation of a manganese oxide coating on filter media. The formation of this coating is an essential prerequisite for efficient manganese removal. There are several types of manganese oxides, which have different affinities for autocatalytic adsorption/oxidation of dissolved manganese. The aim of this study was to characterize the manganese oxide(s) present on filter media from successfully operating manganese removal plants. Characterization of filter media samples from full-scale groundwater treatment plants and identification of manganese species was carried out by X-ray diffraction (XRD), Scanning Electron Microscopy coupled with Energy Dispersive X-radiation (SEM-EDX), Raman spectroscopy and Electron Paramagnetic Resonance (EPR).

The Raman spectroscopy, XRD and SEM analyses showed that the manganese oxide in the coating of the manganese removing filter media is poorly crystalline. Raman spectroscopy and EPR analysis showed that the predominant manganese oxide, responsible for an effective removal of dissolved manganese, is of a Birnessite type. Calculation of ΔH and the g factor from EPR analysis and comparison of these parameters with results from literature identified the Birnessite to be of physicochemical origin, but the sampling after a ripening period of about 15 years does not exclude that Birnessite formation starts via a biological pathway. It is generally accepted that the manganese oxidation pathway is via Hausmannite and Manganite. However, the results transpiring from this research show that in water treatment practice, oxidation of manganese on the surface of manganese removal filter media causes formation of a Birnessite type of manganese oxide. Birnessite has very good properties for adsorption and autocatalytic oxidation of dissolved manganese.

The knowledge that Birnessite is the predominant manganese oxide in filter media effectively removing manganese, may enable shortening of the ripening time in conventional aeration-filtration groundwater treatment plants by creating conditions that favor the formation of this compound. Another possibility is using freshly coated filter media containing Birnessitte to enhance the start of the filter media ripening process. Therefore it is necessary to install the freshly produced Birnessite containing filter media at a filter height where manganese removal takes place.

8.4 Biological and physico-chemical formation of Birnessite during ripening of manganese removal filters

The efficiency of manganese removal in conventional groundwater treatment consisting of aeration followed by rapid sand filtration, strongly depends on the ability of filter media to promote adsorption of dissolved manganese and its subsequent autocatalytic oxidation. Parallel studies, described in this thesis, have shown that the compound responsible for the autocatalytic activity in ripened filters is a manganese oxide called Birnessite. The aim of this study was to determine if the ripening of manganese removal filters and the formation of Birnessite on virgin sand is initiated biologically or physico-chemically. The ripening of virgin filter media in a pilot filter column fed by pre-treated manganese containing groundwater was studied for approximately 600 days. Samples of filter media were taken at regular time intervals, and the manganese oxides formed in the coating were analysed by Raman spectroscopy, Electron Paramagnetic Resonance (EPR) and Scanning Electron Microscopy (SEM).

These analyses confirmed that during the whole period of filter media ripening, a Birnessite type of manganese oxide was the predominant mineral in the coating. Furthermore, Raman spectroscopy results showed that Birnessite was already present in the coating at a very early stage of the ripening process. EPR analyses and comparison with literature showed that the Birnessite type of manganese oxide, at the beginning of the ripening process was of biological origin. With the progress of filter ripening and development of the coating, Birnessite formation became predominantly of physico-chemical origin, although biological manganese oxidation continued to contribute to the overall manganese removal. Especially, solids collected from filter backwash water throughout the whole ripening period were consistently of biological origin, suggesting that biological oxidation of adsorbed manganese was present throughout the filter run. Finally SEM micrographs, showed a clear difference between biologically and physico-chemically formed Birnessite. Biologically produced Birnessite is fluffy, plate structured, whereas physico-chemically produced Birnessite shows a sponge or coral structure.

Figure 8.1: *Sponge or coral structure of physico-chemically produced Birnessite (magnification 10,000 x)- Photo made by Arie Zwijnenburg (Wetsus).*

The knowledge that manganese removal in conventional groundwater treatment is initiated biologically, may help reducing long ripening times by creating conditions that are favourable for the growth of manganese oxidizing bacteria, *e.g.,* by limiting back wash frequency and / or intensity. In practice back wash frequency can be limited, by reducing the iron loading on the filter. Possibilities to reduce iron loading in practice, are lowering the filtration rate, partially circulating of filtrate and or feeding the filters with water containing low amounts of Fe^{2+} (if available).

127

8.5 Identification of the bacterial population in manganese removal filters

Fast filter media ripening, for manganese removal in conventional aeration-rapid sand filtration groundwater treatment, greatly depends on the way autocatalytic adsorptive MnO_x is formed. As said before, the oxide responsible for fast autocatalytic action in ripened filters is (biological formed) Birnessite. The aim of the study described in this chapter was to identify the bacteria present in recently ripened manganese removal filters. For this purpose molecular DNA analysis, such as "next generation DNA sequencing", qPCR and MALDI-TOF analysis were used.

Based on the results of this study, especially from "next generation DNA sequencing" analyses, a bacteria population shift was established, from the iron oxidizing genus *Gallionella*, found in the iron removal filter to the manganese and nitrite oxidizing genus *Pseudomonas* and *Nitrospira*, respectively, present in the manganese removal filter. However, 47.6% of the bacteria population in the manganese oxidizing column, belongs to smaller populations or could not be identified. Applying qPCR it was shown that the most abundant manganese oxidizing genus was *Pseudomonas* sp. Furthermore it was established that the presence of the well-known Mn^{2+}-oxidizing species *Pseudomonas putida* was very limited. Less than 0.01% of the genus *Pseudomonas* present, was of the species *Pseudomonas putida*.

At GWTP Grobbendonk, *Pseudomonas* sp. is most likely the manganese oxidizing bacterium genus playing an important role in initiating filter media ripening However, it is not known whether this bacterium genus is operating alone or as part of a microbial consortium.

With MALDI-TOF analysis, after successive culturing, some *Pseudomonas* species, were identified, amongst others: *P. gessardii, P. grimontii and P. koreensis.*

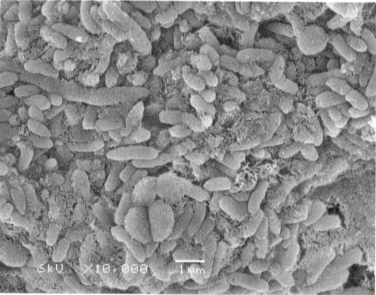

Figure 8.2: *Pseudomonas koreensis (magnification 10,000 x)- Photo made by Jelmer Dijkstra (Wetsus), Sample preparation by Pim Willemse (WLN)*

However, isolated species of *Pseudomonas (grimontii and koreensis)* tested in a fermentor were not able to produce MnO_x under the performed conditions, whereas a laboratory species of *P. putida* was able to do so. This strengthens the hypothesis that in practice not just one bacterium is responsible for the oxidation of manganese, but a *microbial consortium*.

The initiation of manganese removal in conventional groundwater treatment by bacteria, suggests that the ripening time of manganese removing filters could be substantially reduced by creating conditions favorable for these bacteria to oxidize manganese, as mentioned already in section 8.4. Experiments to inoculate filters with manganese oxidizing bacteria cultures to enhance filter media ripening may support this proposal.

8.6 Reduction of ripening time of full scale manganese removal filters with manganese oxide coated media

To achieve effective manganese removal, by conventional aeration filtration with virgin filter media requires a long ripening time. The aim of this study was to assess the potential of manganese oxide-coated filter media to reduce the ripening time of filters with virgin media, under practical conditions. Two full scale filters with virgin sand and anthracite/sand were operated at two groundwater treatment plants, in parallel with full scale test filters, with an additional layer of 0.1-0.3 m of Manganese Oxide-Coated Sand (MOCS) or Manganese Oxide-Coated Anthracite (MOCA).

The ripening time required to achieve complete manganese removal in the full scale filters with virgin sand and virgin anthracite/sand filter media at the two GWTPs; De Punt (The Netherlands) and Grobbendonk (Belgium) was 55 and 16 days, respectively.

The observed differences could be attributed to different feed water quality (pH, redox potential, Fe^{2+}, NH_4^+), applied process design and operational conditions, such as: different iron loading, and backwashing intensity and frequency.

In batch experiments, both fresh MOCA and MOCS showed good manganese adsorptive properties. Addition of a shallow layer of fresh MOCA in test filters eliminated the ripening time completely, while a shallow layer of dried MOCS introduced to virgin sand filters did not significantly reduce the ripening period. The poor performance of MOCS was likely caused by the use of dried MOCS that had lost its adsorption capacity and biological activity.

Applying freshly coated MOCS and MOCA filter media containing the highly adsorptive Birnessite, which is auto catalytically and biologically active in the zone where manganese removal normally takes place offers the opportunity to enhance the start of filter media ripening.

8.7 Factors controlling the ripening of manganese removal filters in conventional aeration-filtration groundwater treatment

It is known from practice that in conventional aeration-filtration groundwater treatment, water quality and operational parameters are very important. The use of manganese oxide coated filter media was found to be beneficial, playing an important role in enhancing/speeding up filter media ripening.

The aim of this research was to study the effect of backwashing and type of virgin filter media on the duration of the ripening period required to achieve effective manganese removal. In addition, the potential of fresh and dried sand and anthracite manganese coated media (MOCS/MOCA) to reduce the ripening period with virgin media was investigated.

This study confirmed that the backwash frequency and therefore also iron loading are important factors regarding the start of filter media ripening, concerning manganese removal. Filter backwashing has a negative influence on the filter media ripening time with virgin filter media. In addition, water quality (high iron concentration) has a negative effect on the filter media ripening, because high iron concentration in the feed water results in a high iron loading of the filter thereby causing the need for frequent backwashing.

As presented in a former section of this chapter, in a pilot study carried out under controlled conditions, freshly prepared manganese oxide coated filter media showed excellent properties to enhance the ripening process, substantially decreasing the ripening time of the filter media. On the contrary, the use of dried MOCS which can only temporarily adsorb Mn^{2+}, did not have a significant impact on ripening time. Moreover, manganese removal efficiency and adsorptive capacity depend on the quality and amount of Birnessite present. From this study it was also concluded that, under comparable conditions, filter media ripening with virgin sand and virgin anthracite, were similar.

Knowledge about the effect of backwashing on filter media ripening and the impact of iron loading on backwash frequency, and the fact that freshly prepared coated media must be used, can be very helpful in reducing ripening time. Especially limiting the backwash frequency at the start of filter ripening with virgin filter media, which will enhance the growth of the bacteria population able to oxidize Mn^{2+} and consequently form Birnessite, offers good opportunities for enhanced filter media ripening.

This study confirmed the findings from section 8.6, that the use of freshly coated filter media could completely eliminate filter media ripening, thus achieving complete manganese removal even at the start up of a virgin filter. Furthermore, the use of freshly coated filter media, in combination with virgin filter media, was not negatively influenced by filter backwashing, manifesting complete manganese removal from the start. Therefore the use of fresh coated filter media will enhance filter media ripening, as concluded before in section 8.6.

8.8 General outlook, limitations and recommendations

The outcome and conclusions of this research provide a better understanding of the mechanisms involved in ripening virgin filter media in traditional aeration-rapid sand filtration systems:

- Firstly, via an inventory of water quality and operational parameters with respect to complete manganese removal, in combination with iron and ammonia removal in the same filtration step, were defined.
- Furthermore, Birnessite was identified as the manganese oxide present in filters that performed well in terms of manganese removal. This highly reactive manganese oxide is responsible for fast removal of manganese by auto-catalytically adsorption and oxidation. This is contrary to the current theory that Hausmanite is responsible for manganese removal.
- Manganese removal starts with biologically formed Birnessite and as ripening progressed, Birnessite formed was of a more physico-chemical origin.
- *Pseudomonas* sp. is most likely the manganese oxidizing bacterium genus playing an important role in starting up filter media ripening.
- Using freshly prepared MOCS or MOCA, containing Birnessite, can eliminate ripening time completely.
- Filter Backwashing is a key factor controlling the start of filter media ripening in manganese removing filters.

Based on the findings and conclusions, this research offers water companies a solid base to evaluate their water treatment plants, and to determine the feasibility of implementing measures to enhance or speed up filter media ripening.

The obtained knowledge provides water companies insight into developing an innovative manganese removal process to shorten or even completely eliminate ripening of new filters and to prolong the useful life of filter media. This can be achieved by creating optimal conditions for manganese oxidizing bacteria to growth and produce Birnessite *i.e.,* by limiting back wash frequency.

However, there is still some important information not covered in this research, regarding the bacteria involved in the manganese removal process. In particular, we still do not understand why they oxidize manganese and in what way they benefit from it. Furthermore, the role of the bacterial consortium and how they benefit from excretion products of other bacteria (*e.g.,* EPS) is not understood. In a follow up research these aspects, concerning the biology involved, should be further revealed.

Another important option for water companies to optimize filter media ripening is to use fresh Birnessite, and manganese oxidizing bacteria, in coated filter media. In this way, MOCS or MOCA will reduce or eliminate filter media ripening time completely.

Furthermore, results obtained from this research, elucidate the possible success of methods used in the past by water companies to enhance fast filter media ripening (use of "old" filter media, feeding (or seeding) filters with backwash water and recirculate filtrate). Using "old" (*i.e.,* fresh) coated filter media containing sufficient Binessite and manganese oxidizing bacteria, enhanced filter media ripening as explained in this research contrary to dry coated filter media. Feeding or seeding the filters with backwash water was beneficial as manganese oxidizing bacteria and Birnessite were added, enhancing filter media ripening. Finally by recirculating filtrate, iron loading and thus backwash frequency was limited while some bacteria and Birnessite were added by this recirculation process.

List of abbreviations and symbols

ATSDR	-	Agency for Toxic Substances and Disease Registry
BW	-	Back Wash or Back Washing
c_e	-	Equilibrium concentration of the adsorbate (mg/L) - (Freundlich)
c_q	-	The PCR cycle after which the fluorescence signal of the amplified DNA and the probe was detected (threshold cycle)
DNA	-	DeoxyRibonucleic Acid
DOC	-	Dissolved Organic Carbon
EBCT	-	Empty Bed Contact Time
E_h	-	Redox potential
EPR	-	Electron Paramagnetic Resonance
EPS	-	Extracellular Polymeric Substances
EU	-	European Union
FR	-	Filter run
G	-	gauss
GWTP	-	GroundWater Treatment Plant
ΔH	-	Wave length (Electron Paramagnetic Resonance)
hFB	-	Filter bed depth (in m) in PCA analysis
ICP-MS	-	Inductive coupled plasm-mass spectrometry
IOCS	-	Iron Oxide Coated Sand
IMnI	-	International Manganese Institute
K	-	Isotherm constant (Freundlich)
KMO	-	Kaiser-Meyer-Olkin statistic test
MALDITOF	-	Matrix-Assisted Laser Desorption/Ionization Time Of Flight Mass spectrometry
MOCA	-	Manganese Oxide Coated Anthracite
MOCS	-	Manganese Oxide Coated Sand
MnO_x	-	Manganese Oxide(s)
N	-	Isotherm constant (Freundlich)
NOM	-	Natural Organic Matter
ORP	-	Oxidation-Reduction Potential
PCA	-	Principal Component Analysis
$p\varepsilon$	-	Electron activity
q_e	-	amount of adsorbate adsorbed per unit mass of the adsorbent (mg/g) - Freundlich
pH_{PZC}	-	pH of zero point of charge
qPCR	-	quantitative Polymerase Chain Reaction
RIVM	-	Rijks Instituut voor Volksgezondheid en Milieuhygiëne
SEM-EDX	-	Scanning Electron Microscopy - Energy dispersive X-Ray Spectroscopy
tFB	-	EBCT (in minutes) in PCA analysis
TOC	-	Total Organic Carbon
UNEP	-	United Nations Environment Programme
UF	-	Ultrafiltration
USEPA	-	Unites States Environmental Protection Agency
VEWIN	-	VEreneging van drinkWaterbebedrijven In Nederland
Vf	-	Filtration rate in $m^3/m^2.h$
WHO	-	World Health Organization
XRD	-	X-ray diffraction

List of publications and presentations

Publications (peer reviewed journals)

Bruins, J.H., Vries, D. , Petruševski, B., Slokar, Y.M. and Kennedy, M.D. (2014) Assessment of manganese removal from over 100 groundwater treatment plants. Journal of Water Supply: Research and Technology - AQUA, 63(4), 268-280.

Bruins, J.H., Petruševski, B., Slokar, Y.M.,Huysman, K., Joris, K., Kruithof, J.C. and Kennedy, M.D. (2015) Biological and physicochemical formation of Birnessite during the ripening of manganese removal filters. Water Research, 69 (C), 154-161.

Bruins, J.H., Petruševski, B., Slokar, Y.M.,Kruithof, J.C. and Kennedy, M.D. (2015) Manganese removal from groundwater: characterization of filter media coating. Desalination and Water Treatment, 55 (7), 1851 – 1863.

Bruins, J.H., Petruševski, B., Slokar, Y.M.,Huysman, K., Joris, K., Kruithof, J.C. and Kennedy, M.D. (2015) Reduction of ripening time of full-scale manganese removal filters with manganese oxide-coated media. Journal of Water Supply: Research and Technology - AQUA, 64(4), 434-441.

Bruins, J.H., Petruševski, B., Slokar, Y.M.,Huysman, K., Joris, K., Kruithof, J.C. and Kennedy, M.D. (2016) Factors controlling the ripening of manganese removal filters in conventional aeration-filtration groundwater treatment. Submitted to Desalination & Water Treatment.

Bruins, J.H., Petruševski, B., Slokar, Y.M., Wübbels, G.H., Huysman, K., Joris, K., Wullings, B., Kruithof, J.C. and Kennedy, M.D. (2016). Identification of the bacterial population in manganese removal filters. Submitted to Water Science and Technology: Water Supply.

Conference presentations

Bruins, J.H., Petruševski, B., Slokar, Y.M. and Kennedy, M.D. (2015) The importance of bacteria for sustainable manganese removal. *AWWA Water Quality Technology Conference 2015* (15-19 Nov), Salt Lake City (UT), USA.

Bruins, J.H., Petruševski, B., Slokar, Y.M., Huysman, K., Joris, K., Kruithof, J.C. and Kennedy, M.D. (2015) Biological and physico-chemical formation of Birnessite during the ripening of manganese removal filters. BioGeo colloquium - Friedrich Schiller University June 30th, Jena, Germany.

Bruins, J.H., Petruševski, B., Slokar, Y.M. and Kennedy, M.D. (2014) Sustainable manganese removal from groundwater through aeration-rapid sand filtration: advantages and problems. *AWWA Water Quality Technology Conference 2014* (16-20 Nov), New Orleans (LA), USA.

Wubbels, G.H., **Bruins, J.H.,** Bosman M. and Woerdt van der D. (2014) The 5th International Slow Sand and Alternative Biological Filtration Conference (19-21 June), Nagoya, Japan.

Bruins, J.H., Petruševski, B., Slokar, Y.M. and Kennedy, M.D. (2013) The use of MOCS and MOCA to shorten ripening time of filter media for manganese removal from groundwater. *2013 inorganic contaminants symposium (5-6 Feb), Sacramento (California), USA*

Bruins, J.H., Petruševski, B., Slokar, Y.M. and Kennedy, M.D. (2012) Characterization and identification of manganese oxides present in naturally coated filter media from conventional aeration and-filtration groundwater treatment plants. *The 4th IWA Asia-Pacific Young Water Professionals Conference 2012* (7-10 Dec), Tokyo, Japan.

Bruins, J.H., Petruševski, B., Slokar, Y.M. and Kennedy, M.D. (2011) Critical review of manganese removal from groundwater: an overview of 100 manganese removal treatment plants. *IWA groundwater specialist conference 2011* (8-10 Sep), Belgrade, Serbia.

Maas van der Maas, P.M.F., Woerdt van der D., **Bruins J.H.** (2009) Effect of Low Pressure UV on the Regrowth Potential of Drinking Water. 5th international congress on ultraviolet technologies (21-23 Sep), Amsterdam, The Netherlands.

Publications (professional papers) / reports

Hofman-Caris, R., Hofs, B., Vries, D., **Bruins, J.H.** (2013), Knowledge inventory manganese removal (in Dutch), BTO-report no. 2013.018.

Huysman, K., Joris, K., **Bruins, J.H.** (2011) Fast filter media ripening (nitrification) in RSF (in Dutch). H$_2$O, 44(5), 55 - 57.

Bruins, J.H., Heeroma, A., Dost, S. (2009) Milk of lime in hardness reduction (in Dutch). H$_2$O, 42(3), 10 - 11.

Maas van der Maas, P.M.F., Woerdt van der D., **Bruins J.H.,** Kooij van der D. (2009) Effect of Low Pressure UV on the Regrowth Potential of Drinking Water (in Dutch). H$_2$O, 42(18), 10 - 11.

Curriculum Vitae Jantinus Bruins

Jantinus Bruins was born in a small village called Eext, in the northern part of The Netherlands, on the 11th of January 1961. After he finished the laboratory school in 1979, he started to work at "Zuiveringsschap Drenthe", a water board, as laboratory assistant. After ten years (1989) he found a job as a laboratory assistant and later as water treatment employee at "het Gemeentelijk Waterbedrijf Groningen - GWG", a municipal drinking water production company. In the meantime he obtained his BSc in Environmental Technology (part time study). In 1998, he joined the new company "Waterbedrijf Groningen", as water technology specialist after the merger of GWG with the drinking water company of the province of Groningen. Also he studied (part time) Environmental Science from 1999-2002. He obtained his Msc in 2002. In 2000 he started to work for WLN, the water quality and technology advice center for two Dutch water companies in the northern part of The Netherlands, as senior technology advisor. Jantinus is an experienced specialist in water treatment, with more than 35 years work experience in his field. His working field covers (environmental) technology for drinking- and industrial water supplies and wastewater treatment, with special emphasis on drinking water technology.

Printed and bound by CPI Group (UK) Ltd, Croydon, CR0 4YY

18/10/2024

01776249-0001